Christopher Friedrich Reiche
Novel sensors for scanning force microscopy based on carbon nanotube mechanical resonators

TUDpress

DRESDNER BEITRÄGE ZUR SENSORIK

Herausgegeben von

Gerald Gerlach

Band 62

Christopher Friedrich Reiche

Novel sensors for scanning force microscopy based on carbon nanotube mechanical resonators

TUDpress

2016

Die vorliegende Arbeit wurde unter dem Titel „Novel sensors for scanning force microscopy based on carbon nanotube mechanical resonators“ am 01.10.2015 als Dissertation an der Fakultät Mathematik und Naturwissenschaften der Technischen Universität Dresden eingereicht und am 20.04.2016 verteidigt.

Gutachter: Prof. Dr. Bernd Büchner, Technische Universität Dresden
Prof. Dr. Georg Weiß, Karlsruher Institut für Technologie

Bibliografische Information der Deutschen Nationalbibliothek
Die Deutsche Nationalbibliothek verzeichnet diese Publikation in der Deutschen Nationalbibliografie; detaillierte bibliografische Daten sind im Internet über http://dnb.d-nb.de abrufbar.

Bibliographic information published by the Deutsche Nationalbibliothek
The Deutsche Nationalbibliothek lists this publication in the Deutsche Nationalbibliografie; detailed bibliographic data are available in the Internet at http://dnb.d-nb.de.

ISBN 978-3-95908-068-2

Eckhard Richter & Co. OHG
Bergstr. 70 | D-01069 Dresden
Tel.: 0351/47 96 97 20 | Fax: 0351/47 96 08 19
http://www.tudpress.de

TUDpress ist ein Imprint von w. e. b.

Gesetzt vom Autor.
Printed in Germany.

Editor's Preface

Scanning force microscopy (SFM) is a very unique and versatile technique to probe surfaces with a resolution on the order of parts of a nanometer, i.e., several orders of magnitude better than the optical diffraction limit. The functional principle is based on the coupling force between the mechanical probe tip and the surface. Its tremendous sensitivity with respect to small forces can be used to exploit the SFM principle for other actions of forces like that one originating from very small magnetic sources.

This work presented here uses dynamic-mode SFM for the quantitative measurement of very small magnetic effects by using iron-filled carbon nanotubes as probe tips. The particular design of the probe allows to use even two flexural modes to determine both the out-of-plane and the in-plane force gradient components.

To further increase the sensitivity of the SFM probe the set-up is modified to correspond to a co-resonant oscillator: by adjusting the resonance frequency of the probe tip to that one of a second resonator a significant amplification of the measurement signal can be achieved. These modified probes then use for the very first time resonant oscillations of iron-filled carbon nanotubes in SFM.

Since this new technique seems to open up unimagined possibilities for the future development of measuring very small magnetic quantities, I feel confident that this volume of the book series 'Dresden Contributions to Sensor Technology' will earn the amount of attention it deserves.

Dresden, May 2016 Gerald Gerlach

Kurzzusammenfassung

Rasterkraftmikroskopie im dynamischen Betriebsmodus ist eine vielseitige Untersuchungsmethode der Oberflächenphysik, mit der Kräfte und kraftbezogene Eigenschaften nahe einer Probenoberfläche mit Nanometerauflösung gemessen werden können. Vor kurzem wurden neuartige bidirektionale Rasterkraftmikroskopiesensoren und zugehörige Betriebsmodi eingeführt. Diese Sensoren erlauben Messungen von Kraftkomponenten beziehungsweise deren Ableitungen sowohl entlang der senkrechten als auch der waagerechten Richtung. Die gewünschte Richtungskomponente ergibt sich allein durch Anregung entsprechender Biegeschwingungsordnungen der den Sensoren zugrunde liegenden Cantilever.

Im Rahmen dieser Dissertation wird die Anwendbarkeit dieser neuartigen Sensoren auf die quantitativen Messung von Kraftgradienten erweitert. Zu diesem Zweck wurden die effektiven Federkonstanten der Sensoren für beide Messrichtungen abgeleitet. Die ensprechenden mathematischen Ausdrücke wurden experimentell durch die Messung von magnetostatischen Kräften verifiziert. Außerdem gelang es, eine Erweiterung des Meßprinzips auf drei orthogonale Messrichtungen zu demonstrieren. Für diese Experimente wurden die Sensoren mit speziellen magnetischen Messspitzen ausgerüstet: mit eisengefüllten Kohlenstoffnanoröhren.

Die eisengefüllten Kohlenstoffnanoröhren wurden ferner als Testprobe für Magnetometer mit Nanometerabmessungen angewendet, um deren Anwendbarkeit für die Untersuchung magnetischer Nanopartikel zu zeigen.

In dieser Arbeit wird außerdem ein neuartiges ko-resonantes Sensorkonzept eingeführt, welches auf dem Modell eines asymmetrischen gekoppelten harmonischen Oszillatorsystems basiert. Durch eine mechanische Kopplung eines leicht messbaren mikroskaligen Oszillators mit einem nanoskaligen Oszillator, der eine niedrige Federkonstante aufweist, können neuartige Sensoren mit stark erhöhtem Messsignal hergestellt werden. Entscheidend ist hierbei, dass die Resonanzfrequenzen der beiden Einzeloszillatoren aufeinander abgestimmt werden.

Dieses generell anwendbare Konzept wurde dann auf die bidirektionalen Kraftmikroskopiesensoren angewandt, um diese erheblich zu verbessern. Dazu wurde die Resonanzfrequenz der eisengefüllten Kohlenstoffnanoröhre mittels elektronenstrahlgestützter Materialabscheidung derart angepasst, dass sie nahe bei der Resonanzfrequenz des waagrechten

Betriebsmodus liegt. Da der Verstärkungsfaktor experimentell bestimmt oder anhand der Resonanzfrequenzen und Federkonstanten der Teilsysteme berechnet werden kann, sind Sensoren die auf diesem ko-resonanten Konzept basieren, ebenfalls für quantitative Messungen anwendbar. In entsprechenden Experimenten wurde eine Signalverstärkung von mehr als drei Größenordnungen erreicht. Nach dem Kenntnisstand des Autors ist dies außerdem die erste Rasterkraftmikroskopiemessung, die auf der Biegeschwingung einer Kohlenstoffnanoröhre beruht.

Abstract

Dynamic mode scanning force microscopy is a versatile method for studying forces and force-related properties close to a sample's surface with sub-nanometer resolution. Recently, a novel sensor concept for bidirectional scanning force microscopy and corresponding modes of operation have been introduced that support the measurement of perpendicular and lateral force components or their derivatives, respectively. This sensor concept is based on the excitation of the first and second flexural oscillation mode of a cantilever beam and corresponding sensors are compatible with standard scanning force microscopy equipment.

In the framework of this thesis the applicability of these sensors is advanced to quantitative scanning force microscopy by deriving the dynamic effective spring constants for both measurement directions. The corresponding expressions were validated successfully by magnetostatic force gradient measurements. Additionally, it was possible to demonstrate an expansion of the operating principle of the sensors to three orthogonal directions. In these experiments the bidirectional sensors were equipped with very special magnetic interaction tips to facilitate the measurements: iron-filled carbon nanotubes.

The iron filled carbon nanotubes were also applied as test samples to demonstrate the viability of nanoscaled magnetometer devices for studying magnetic nanoparticles.

In addition to these results a novel co-resonant sensor concept is proposed, based on theoretical considerations on an asymmetric coupled harmonic oscillator model: By mechanically coupling an easy-to-detect microscaled oscillator with a low spring constant nanomechanical oscillator and matching the resonance frequencies of these subsystems it is possible to create a new class of mechanical oscillation-based force- and mass sensors with a greatly enhanced measurement signal.

This general result was employed to improve the bidirectional sensors: the fundamental resonance frequency of the iron-filled carbon nanotube interaction tip was adjusted to be close to that of the lateral operation mode of the sensor. It was also demonstrated that the signal amplification can be experimentally calibrated or calculated based on the parameters of the uncoupled subsystems. Therefore, sensors produced according to this co-resonant concept are still viable for quantitative measurements. The experimental results include a signal enhancement by more than three orders of magnitude compared to the expected values for an unmodified sensor. To the author's knowledge this is also the first demonstration of scanning force microscopy based on an oscillating carbon nanotube.

Contents

List of abbreviations

BE	back-scattered electrons
CNT	carbon nanotube
dSFM	dynamic mode scanning force microscopy
FeCNT	iron-filled carbon nanotube
FIB	focused ion beam
LFM	lateral force microscopy
MFM	magnetic force microscopy
MWCNT	multi-walled carbon nanotube
nano-SQUID	nanoscaled superconducting quantum interference device
PLL	phase-locked loop
PE	primary beam electrons
Q factor	quality factor
SE	secondary electrons
SEM	scanning electron microscopy
SFM	scanning force microscopy
SWCNT	single-walled carbon nanotube
SQUID	superconducting quantum interference device
STM	scanning tunneling microscope

1 Introduction

A tiny piece of gold foil and a diamond with a sharp tip: these raw materials formed the first scanning force microscopy force sensor, an invention which sparked a whole new field of surface science.

After his contribution to the scanning tunneling microscope (STM)[1], Gerd K. Binnig invented the atomic force microscope [2], which was based on similar principles and presented his first experimental results together with his coworkers to the public nearly three decades ago in 1986 [3]. Their first implementation of this new type of microscope used an STM to measure the deflection of a force sensing element consisting of a gold foil cantilever beam and a diamond interaction tip in order to probe forces close to a sample surface. In this first publication they introduced the atomic force microscope as a method to investigate inter-atomic forces with very high spatial resolution on conductors as well as on insulators. The ability to study the latter was an improvement over the STM, which is limited to conductive samples only. Binnig et al. had already suggested the generality of their principle with respect to all forces and proposed many possible operation modes: static and dynamic, contact and non-contact.

Since this first publication nearly every aspect of the method has been enhanced. Frequently, the more generic term scanning force microscopy (SFM) is used to stress the applicability to many different force-types. These include the already mentioned inter-atomic but also electrostatic and magnetic forces. In modern SFM instrumentation, the STM detection has been replaced by optical methods or in some cases even more complex measurement systems, such as piezoelectric detection elements integrated directly into the force sensor. The force sensors have also been improved and are nowadays mass-produced by microfabrication techniques, with a huge range of different geometries as well as material and tip combinations available. These sensors can be utilized in many different operation modes, which have been developed over the years since the introduction of the method.

One of the basic operating principles of SFM is the dynamic mode. In this mode the force sensor is mechanically excited at or close to one of its resonance frequencies and force gradients emanating from an interplay between the tip and a sample induce a change of this resonance frequency. This shift or, related to it, the change of the oscillation's amplitude or phase with respect to the excitation are used to derive information on the tip-sample interactions or related properties of the sample.

Two dynamic SFM techniques are of special interest for this work: dynamic magnetic force microscopy (MFM) and dynamic lateral force microscopy (LFM). In MFM, part of the tip material is magnetic, which results in a force sensor that is sensitive to magnetostatic interactions. In dynamic LFM the tip oscillates in a direction lateral to the sample surface. This results in a sensitivity of this mode to lateral instead of the commonly used perpendicular force gradients. This mode can be used to measure friction on the atomic scale or, in combination with MFM, to study lateral magnetic stray field gradients. Even though there are already well established dynamic mode LFM methods, they are either intrinsically limited to the lateral direction or require more complex non-standard dynamic mode SFM (dSFM) equipment. Recently a novel sensor principle for dynamic operation mode MFM has been developed that includes the ability to switch between lateral and perpendicular sensitivity and is compatible with basic dSFM setups [4]. The first proof-of-principle experiments with these new sensors used very special MFM interaction tips to facilitate the measurements: iron-filled carbon nanotubes (FeCNTs).

Carbon nanotubes are tubular structures of carbon with diameters on the nanometer scale. Due to their special structure they exhibit a range of interesting physical properties, such as their high mechanical and chemical stability. This enables many prospective applications, from reinforcement of materials, drug delivery and wear resistant SFM tips with an high aspect ration to nanoscopic force gradient or mass sensors.

In the past, FeCNTs have been shown to have advantageous properties as magnetic tips for MFM [5]. These include their durability owed to the carbon shell as well as a monopole-like stray-field characteristic of the encased iron nanowire. This monopole-like characteristic due to the high aspect ratio of the wire is especially of interest for quantitative measurements because it simplifies the required calculations. Furthermore, the quantity required for those calculations, the monopole-like moment, is independent of the sample's stray field and may be derived from geometric parameters of the FeCNT without the need for a calibration step. Additionally, oscillating FeCNTs were found to be superior magnetostatic force gradient sensors because of their small spring constant [6]. Hence, it would be an attractive prospect to use them in a MFM setup. However, due to their small spatial dimensions the detection of their oscillatory state requires more complex methods than in the case of standard dSFM, which limits their applicability.

The origin of this work was to provide an in-depth study of the above mentioned novel bidirectional MFM method, to experimentally verify its ability for quantitative magnetic field gradient measurements in two directions and to further improve this method. From these starting points, the challenge to improve this method led to a new concept to combine the ease-of-detection of micromechanical resonators with the high force gradient sensitivity of nanoscaled resonator systems. On the one hand, this resulted in a novel class of carbon nanotube resonator-based dSFM sensors, and on the other hand the concept is of a very general nature with many possible applications beyond dSFM.

After this first introductory chapter the remainder of the work is structured as follows: since sensors for dSFM are the main topic of this work, the second chapter focuses on the introduction of dynamic mode scanning force microscopy and related techniques. Subsequently, the experimental techniques to fabricate and characterize the sensors are introduced and discussed in chapter 3. Iron-filled carbon nanotubes were essential building blocks of all the experiments that are presented within this thesis. Therefore, chapter 4 is dedicated to a review of their production, previously studied properties and applications as well as a presentation of new results obtained with this interesting hybrid material system. Chapter 5 deals with the in-depth study of the quantitative bidirectional force gradient microscopy technique that combines standard perpendicular dSFM and LFM. All the formulas necessary for a quantitative evaluation of force gradients from frequency-shift dSFM data are derived and experimentally validated by employing magnetostatic force gradients. As mentioned above, a method to combine at least part of the sensitivity of low-stiffness FeCNTs and standard dSFM equipment would be desirable. In chapter 6 such a method, the co-resonant sensor concept, is derived from a simple asymmetric coupled harmonic oscillators model and its applicability to dSFM and various other mechanical oscillation-based sensors is discussed. Subsequently, this co-resonant detection concept is applied to the bidirectional sensors to create a novel class of high sensitivity dynamic mode LFM sensors. The experimental results with these advanced sensors presented in chapter 7 also provide a verification of the co-resonant detection concept. Finally, chapter 8 summarizes the results and discusses open questions. Furthermore, this last chapter also provides an outlook on future experiments.

2 Scanning force microscopy

This thesis mainly deals with the development and improvement of novel sensors for dynamic mode scanning force microscopy. Therefore, this chapter is dedicated to an introduction in this microscopy method, its various operation modes and approaches to derive quantitative results from measured data. Since the measurements presented in later chapters are based on magnetostatic interactions, a significant amount of this chapter will cover the topic of magnetic force microscopy.

2.1 Introduction to scanning force microscopy

The invention of the scanning force microscope is closely connected to that of the scanning tunneling microscope [3]. Both methods have many advantages and experimental equipment in common; they both offer sub-nanometer spatial resolution, and both employ similar scanning mechanisms and feedback electronics. However, they differ in the means to generate their respective measurement signal. In a STM the signal is generated by measuring the tunnel current between a conductive sample and a sharp metallic tip. In contrast, SFM utilizes a beam-like force sensor equipped with a suitable tip to measure interactions close to a sample surface. An example of such a SFM force sensor, in this case a commercially available microstructured cantilever with a silicon tip, is shown in figure 2.1. One of the main advantages of SFM over using a STM is its applicability to non-conductive samples, including organic samples and oxide surfaces. One of the main challenges of the SFM method stems from the many different types of forces acting on an SFM sensor simultaneously, and the resulting difficulty to derive quantitative information about a specific interaction from the data.

Today there is a great number of different methods based on the SFM principle available. The choice of a specific type of SFM depends on many factors, including the desired force type or material property to be studied as well as the experimental conditions and available equipment. In order to classify the multitude of different SFM methods, various aspects can be taken into consideration. One of those aspects is the operational mode of the force sensor resulting in a differentiation into static and dynamic modes.

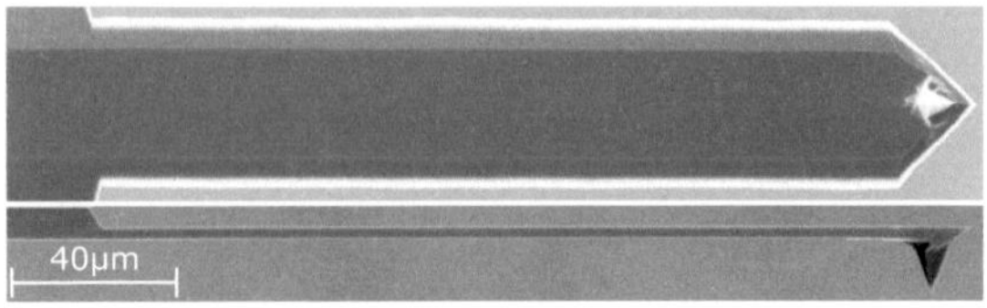

Figure 2.1: Scanning electron microscopy images of a commercially available microstructured scanning force microscopy sensor

In static mode SFM the spring-like cantilever force sensor is used to probe forces between its interaction tip and the sample surface. These forces lead to a static deflection of the cantilever, which can be used as a measurement signal. In dynamic mode SFM, the force sensor is excited to mechanical oscillations and the measurement signal is extracted from changes of the oscillatory state of the sensor caused by elastic and inelastic interactions between tip and sample. [7, pp.18f.]

A further distinction of SFM methods is usually made by considering the distance between interaction tip and sample surface, i.e. if it is in contact with the sample or not. However, the definition of 'contact' and 'non-contact' and their distinction is difficult on atomic length scales and the use of these denominations in literature is ambiguous. A common approach to this problem is to simplify the force acting on a dSFM sensor as an interaction between two neutral atoms. At very small separation between sample and tip atom, the force is repulsive due to Coulomb repulsion between the positive charges of the no longer completely screened nuclei as well as the Pauli repulsion of the electron shells. At larger separations the force is attractive due to contributions from van-der-Waals forces and molecular binding-like interactions [8, p.51]. These contributions can be approximated in terms of Lennard-Jones [9] or Morse [7, p.32] potentials. For larger separations it is possible to improve these simple models by adding the van-der-Waals contribution of the relevant tip and sample volumes [9].

According to these considerations the 'non-contact' label is used when the tip-sample force is attractive with a positive detectable gradient and the 'contact' one refers to an operation in the range of the repulsive tip sample-interactions [10]. The respective ranges are sketched in figure 2.2.

Nevertheless, it is difficult to define these measurement regimes on arbitrary samples as the contributions of various interactions like electro- and magnetostatic forces are not considered in this simple model and those contributions can lead to repulsive forces even far above the sample surface. It has to be noted that especially at early stages of the development of SFM any dynamic mode was considered 'non-contact' and so both

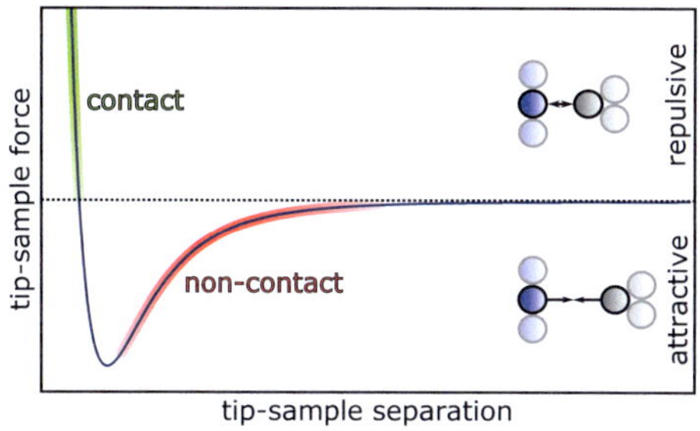

Figure 2.2: Sketch of the tip-sample force derived from a Lennard-Jones potential with long-range van-der-Waals volume contribution. The 'contact' and 'non-contact' regimes are indicated.

denomination were used synonymously [11, p.251][8, p.100]. This is no longer valid, however, since there are dynamic modes, called tapping or intermittent modes, that use comparatively large cantilever amplitudes [12] and can be operated with the tip in the attractive as well as the repulsive regime [13].

2.2 Dynamic mode scanning force microscopy

Dynamic mode scanning force microscopy is a versatile technique, which was demonstrated to be capable of true atomic resolution on many different surfaces [14, 15, 16] and even investigations of subatomic features have been reported [17]. Besides inorganic specimens, dSFM methods can also be used to study organic materials [18] as well as to investigate molecules adsorbed to surfaces [19].

2.2.1 General principle

The general principle of dSFM is relatively simple: a spring-like structure, usually a microfabricated cantilever beam or, in some cases, a quartz tuning fork [20], is equipped with a sharp interaction tip and excited to mechanical oscillations close to or at its fundamental resonance frequency. This combination of tip and cantilever is called 'force sensor'. Interactions between the sample and the tip lead to measurable changes in the sensor's oscillation characteristics determined by the resonance frequency, amplitude and phase with respect to the excitation. The changes to these parameters of the oscillation can be used as a measurement signal. The oscillating force sensor is then moved relative to the surface of a sample in a scanning trajectory to create real space maps of the measurement signal.

A dSFM instrument consists of several main components: the cantilever force sensor with an interaction tip, a means to mechanically excite the sensor, a setup to detect the oscillatory state of the sensor, and a way to approach the sensor to and scan it relative to a sample surface. Usually, at least one feedback-loop is present to change the distance of the sensor to the sample's surface in order to keep a constant value of the tip-sample interaction measurement signal. This can be used to prevent damage to the tip caused by collisions with the sample and to trace the sample's surface topography. However, more sophisticated dSFM methods require additional feedback electronics [21].

The dSFM instrument utilized in this work is based on a very common setup [8, p.45]. It employs microstructured cantilever force sensors [22] that are mechanically excited by a piezoelectric actuator and their oscillatory state is determined by measuring the deflection of a laser beam on the back of the cantilever via a sectioned photo diode. The feedback electronics include loops addressing the sensor-sample distance and both frequency and amplitude of the excitation. A scheme of this setup can be found in figure 2.3.

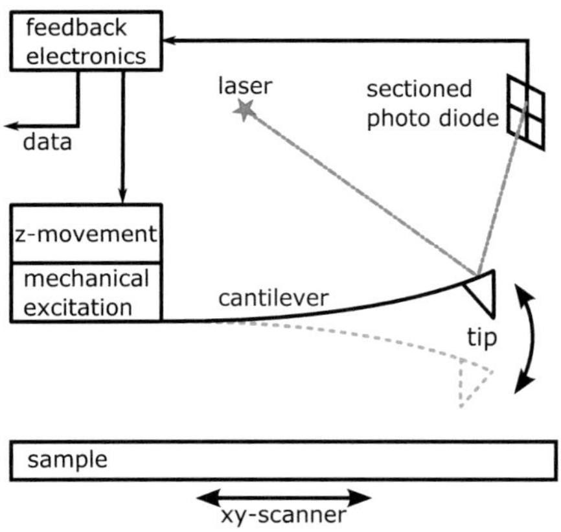

Figure 2.3: Basic dynamic mode scanning force microscopy setup

2.2.2 Harmonic oscillator model of a force sensor

To understand the origin of the measurement signal in dSFM it is instructive to model the cantilever, tip and interaction as a simple driven damped spring-mass harmonic oscillator. This simplification is often used in dSFM [23] and was found to be a good approximation when compared to more complex solutions [24]. The equation of motion of such a harmonic oscillator reads

$$m_{eff}\ddot{s} + d\dot{s} + ks = F_0 cos(\omega_D t) + F(s) \tag{2.1}$$

with the deflection s from the equilibrium position, damping coefficient d, spring constant k and effective mass m_{eff} of the cantilever-tip system. The forces on the right hand side of this equation are separated in the periodic force with amplitude F_0 and angular frequency ω_D exciting the mechanical oscillation of the system and the conservative tip-sample interaction $F(s)$. Please note that the mechanical excitation of the cantilever is often realized by a piezoelectric actuator creating a periodic displacement of the mounting of the cantilever instead of a periodic force. However, this displacement is translated by the spring constant k into a force, so equation 2.1 is also valid for this kind of excitation. Assuming only small deflections, the equation is further simplified by a first order Taylor expansion of the tip-sample interaction near the equilibrium-position $s = 0$ of the free oscillator, which results in:

$$F(s) \approx F(0) + s \left.\frac{\partial F(s)}{\partial s}\right|_{s=0} = F(0) - s\Delta k \tag{2.2}$$

with $\Delta k = -\partial F(s)/\partial s|_{s=0}$. In the following, the short form for the partial derivative with respect to s at the equilibrium position $\partial F(s)/\partial s|_{s=0} = \partial F_s/\partial s$ as well as the simpler term 'force gradient' will be used. Inserting this into equation 2.1 yields the equation of motion for a driven damped harmonic oscillator with a new spring constant $k^+ = k + \Delta k$ and a constant force in addition to the periodic driving force. This constant force $F(0)$ only shifts the equilibrium position of the system by $s_0 = F(0)/k$ [25, p.23] and it was found that with usual dSFM parameters the shift due to this force is very small [26]. Additionally in this linear approximation $F(0)$ does not affect the vibrational parameters of the system and can be neglected in this dSFM model. These simplifications result in an harmonic oscillator model as shown in figure 2.4.

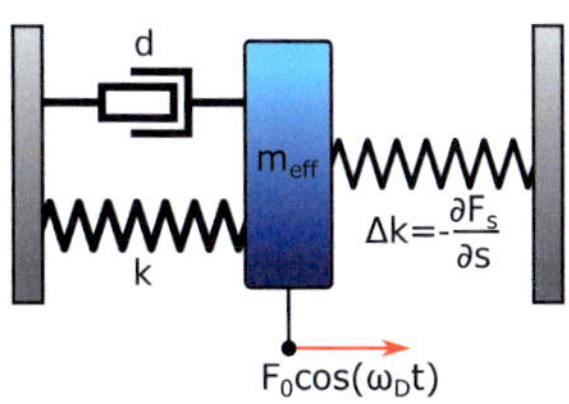

Figure 2.4: Damped harmonic oscillator model of a dynamic mode scanning force microscopy sensor with periodic excitation. The negative first derivative $-\partial F_s/\partial s$ of the conservative tip-sample interaction at the equilibrium position acts as an additional spring Δk. Furthermore, the constant part of the force $F(0)$ was neglected.

The solution to the simplified equation of motion

$$m_{eff}\ddot{s} + d\dot{s} + k^+ s = F_0 cos(\omega_D t) \tag{2.3}$$

for this driven harmonic oscillator can be found in basic textbooks on classical mechanics [27, 28]. This solution is the sum of a particular and a homogeneous solution with the latter describing an exponentially damped motion that vanishes after a sufficient amount of time. In contrast, the particular solution will not vanish due to the periodic driving force and represents the steady-state solution. This steady-state solution is derived as

$$s(t) = Acos(\omega_D t + \varphi). \tag{2.4}$$

By using the angular eigenfrequency

$$\omega_{\Delta k} = \sqrt{\frac{k^+}{m_{eff}}} = \sqrt{\frac{k + \Delta k}{m_{eff}}} \tag{2.5}$$

of a corresponding undamped oscillator with spring constant k^+ and mass m_{eff}, the oscillation amplitude A and phase φ with respect to the periodic excitation are given by the equations:

$$A = \frac{F_0/m_{eff}}{\sqrt{(\omega_D^2 - \omega_{\Delta k}^2)^2 + \omega_D^2 \left(d/m_{eff}\right)^2}} \tag{2.6}$$

$$\tan(\varphi) = \frac{(d/m_{eff})\,\omega_D}{(\omega_D^2 - \omega_{\Delta k}^2)}. \tag{2.7}$$

Here $\varphi \in (-\pi, 0)$ applies. Furthermore, the angular resonance frequency which is given by the angular driving frequency at which the amplitude exhibits a maximum can be derived by solving $\partial A/\partial \omega_D = 0$:

$$\left(\omega_{\Delta k}^*\right)^2 = \omega_{\Delta k}^2 - \frac{1}{2}\left(d/m_{eff}\right)^2. \tag{2.8}$$

Of course we assume that the system is a resonator, which means that is capable of resonant behavior. This requires the right side of equation 2.8 to be positive. Usually when dSFM sensors are considered, the damping is very low, i.e. $\omega_{\Delta k}^2 >> (d/m_{eff})^2$. This results in the angular resonance frequency being approximately equal to the angular eigenfrequency of the corresponding undamped oscillator $\omega_{\Delta k}^* \approx \omega_{\Delta k}$. Therefore, these terms are often used interchangeably in literature concerning dSFM (e.g. [29, 30]). Furthermore, the term 'eigenfrequency' is only well defined for a harmonic oscillator in the absence of damping and describes the frequency of its free oscillation. This frequency is different from that of the decaying free oscillation of the corresponding damped harmonic oscillator, which is also different from its resonance frequency [28, p.92]. The corresponding expression for the eigenfrequency from equation 2.5 is found in the solutions of the damped system, however. Additionally it determines the driving frequency at which the oscillation of the driven damped system exhibits a phase difference with respect to the periodic excitation of $\varphi = -\pi/2$ [28, p.95]. Therefore, this frequency will also be called 'eigenfrequency' of the damped harmonic oscillator in the following.

Please note that in theoretical considerations concerning harmonic oscillators the angular frequency ω is the preferred variable over the standard frequency f, which is usually the accessible physical quantity in dSFM experiments. As both frequencies are related by $\omega = 2\pi f$ and since it is clear from their respective symbols which frequency type is referred to, a distinction in their denomination is unnecessary. Therefore, both will simply be called 'frequency' in the remainder of this thesis.

The damping of the force sensor can be described by its quality factor (Q factor). The Q factor relates the total energy E_{tot} of an oscillation to the energy loss per cycle δE_{cycle} [8, p.81][7, p.398]. It can furthermore be related to the full bandwidth $\delta\omega$ and the resonance frequency $\omega^*_{\Delta k}$ of the oscillation, given a harmonic oscillator and the case of low damping. The full bandwidth is defined as the frequency separation of the two excitation frequencies at which the amplitude of the oscillation has decreased to factor of $1/\sqrt{2}$ of its maximum value [25, p.21]. The Q factor for the harmonic oscillator described by equation 2.3 is given by [31, pp.561-571]

$$Q = 2\pi \frac{E_{tot}}{\delta E_{cycle}} \approx \frac{\omega^*_{\Delta k}}{\delta\omega} \approx \frac{m_{eff}}{d}\omega_{\Delta k} = \frac{\sqrt{k^+ \, m_{eff}}}{d} \tag{2.9}$$

with the second approximation being valid for very low damping. In this context, another useful expression for the harmonic oscillator model can be obtained by inserting the Q factor from equation 2.9 in equation 2.6 for the oscillator's amplitude. For driving frequencies much lower than the resonance frequency of the oscillator its amplitude is given by $A_0 = F/k$ [32]. The Q factor then provides a simple equation for the amplification of this amplitude by the resonance:

$$\frac{A(\omega_D)}{A_0} = \frac{\omega^2_{\Delta k}}{\sqrt{(\omega^2_D - \omega^2_{\Delta k})^2 + \frac{\omega^2_D \omega^2_{\Delta k}}{Q^2}}}. \tag{2.10}$$

In the case of an excitation at the resonance frequency this reduces to $A^* = Q \cdot A_0$ [8, p.93].

With the above equations the basic measurement principle of non-contact dSFM can be understood. A force gradient acting on the sensor's tip may be regarded as an additional spring which shifts the system's eigenfrequency $\omega_{\Delta k}$ resulting in a change of its amplitude and phase versus driving frequency curves. The dSFM signal is then generated by tracking these changes by various methods. The most commonly employed of these methods are discussed in the following.

2.2.3 Amplitude- and frequency modulation

A comparatively simple method is called 'amplitude modulation' or sometimes 'slope detection'. Its basic principle has already been proposed by Binnig et al. in their first paper [3] and its experimental use was pioneered by Martin et al. [33]. According to this method the system is excited at a constant driving frequency ω_D^{const} above or below the resonance frequency where the slope of the amplitude versus driving frequency curve is large. A force gradient induced change of the eigenfrequency will strongly change the amplitude of the system at the given fixed excitation frequency according to equation 2.6. This change in amplitude can then be used as the measurement signal. The minimal detectable force gradient was found to be proportional to $(\sqrt{Q})^{-1}$, so sensors with a high Q factor, i.e. low damping would be desirable in order to achieve maximum sensitivity [8, p.93]. Any amplitude change, however, will trigger a transient state of the system and the required time to reach the steady state also increases with decreasing damping, limiting the practical use of high Q factor sensors. Furthermore, any additional external damping on the force sensor will also result in an amplitude decrease that is indistinguishable from one induced by a force gradient.

These problems were solved by the 'frequency modulation' method proposed by Albrecht et al.[21]. By using additional feedback electronics to keep the oscillation phase fixed at $\varphi = -\pi/2$ the driving frequency is shifted to always keep the system excited at its eigenfrequency. The frequency shift

$$\Delta\omega = \omega_{\Delta k} - \omega_0 \approx \frac{\omega_0}{2k}\Delta k = -\frac{\omega_0}{2k}\frac{\partial F_s}{\partial s} \tag{2.11}$$

compared to the force gradient free case is then used as the measurement observable. The approximation in equation 2.11 assumes $k >> |\Delta k|$, a condition that is usually well satisfied in dSFM [23]. With this method the oscillation amplitude no longer carries the information about the force gradients and another feedback loop can be used to keep it constant throughout the measurement, preventing transient states and thus increasing measurement speed. Furthermore, for a high Q factor force sensor the primary origin of a reduction of the oscillation amplitude is external dissipation and the change of excitation amplitude needed to keep oscillation amplitude constant can be used to monitor this dissipation [8, p.84]. Two additional feedback electronics are needed for this 'frequency modulation' method, however, resulting in an increased complexity of the equipment and its operation. It is emphasized that while equation 2.11 includes the eigenfrequency ω_0, usually the resonance frequency of the oscillator is the experimentally more accessible quantity. As mentioned above, the eigenfrequency can be well approximated by the resonance frequency in the case of low damping. This approximation will always be employed if equation 2.11 or derived equations are used in the experimental parts of this thesis.

The 'amplitude modulation' method is usually used in dSFM experiments at ambient pressure or in liquids, since damping from the cantilevers movement in a medium results in low Q factors, resulting in an appropriate data acquisition speed without the need for additional feedback loops. In contrast, 'frequency modulation' is utilized in vacuum conditions where large Q factors are common [30]. A visualization of both methods is provided in figure 2.5. It has to be noted that sometimes in literature the terms 'amplitude modulation' mode and 'tapping' mode are used synonymously. However, the second one is a subset of the first [34].

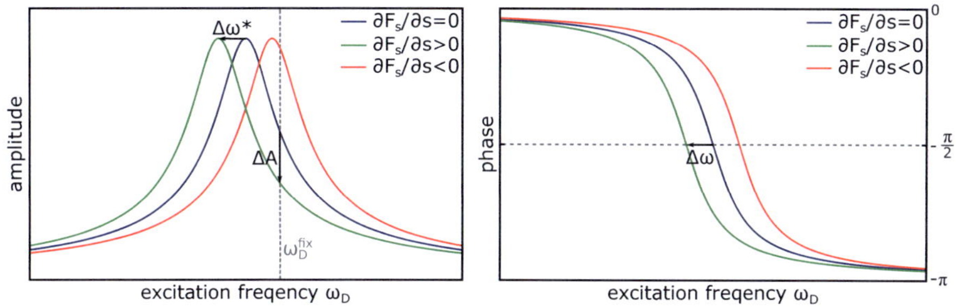

Figure 2.5: Amplitude and phase versus driving frequency curves of the discussed harmonic oscillator model. A positive force gradient acting on the oscillator induces a negative shift $\Delta\omega$ of the eigenfrequency of the oscillator and vice versa. The measurement signals acquired according to the amplitude and frequency modulation methods are indicated (ΔA and $\Delta\omega$ respectively) for the case of a positive force gradient. In the common case of low damping the shift of the resonance frequency $\Delta\omega^*$ is approximately equal to the shift $\Delta\omega$ of the eigenfrequency.

These considerations regarding the tip-sample force gradient as an additional spring are valid as long as the assumption of an approximately constant force gradient $\partial F(s)/\partial s$ over the whole oscillation period holds. For long range forces like electro- and magnetostatic interactions the characteristic length scale for changes is usually far bigger than the usual operation amplitude of the sensor, thus this assumption is justified. In the case of large oscillation amplitudes or short range inter-atomic forces, however, which change by several orders of magnitude over the range of the tip movement, the relationship between the frequency shift and the forces is more complex [9]. In order to calculate the local tip-sample force from a frequency shift signal, several sophisticated approaches are available which involve matrix methods [35] or the integration of quantities derived from the frequency shift signal and the oscillation amplitude of the dSFM sensor [36]. An experimental application of the latter method has been demonstrated by reconstructing the forces acting on a single cobalt atom on platinum surface [37].

Real dSFM is of course not limited to one dimension as the model above and both the sensor movement and the measured force have three spatial components. For an oscillation of the sensor in the direction of an unit vector $\vec{n}$ the force gradient that changes the cantilever's resonance frequency is given by [38]

$$\frac{dF_n}{dn} = \vec{n} \cdot \vec{\nabla} F_n = \vec{n} \cdot \vec{\nabla} \left(\vec{n} \cdot \vec{F}(\vec{r}) \right). \tag{2.12}$$

In the special case of a unit vector along one of the basic directions $s = x, y, z$ this reduces to $\partial F_s / \partial s$.

2.3 Cantilever beams

While the simple harmonic oscillator model provides insight in the signal formation in dSFM, more complex methods are required to calculate the dynamic behavior and relevant parameters of a dSFM force sensor. The obtained results can then be transferred to the simple harmonic oscillator model. As mentioned before, a typical dSFM sensor consists of a spring-like cantilever beam and a sharp tip. To begin with, the tip is neglected and the static and dynamic mechanical properties of a cantilever beam with constant cross section A along the cantilever's length L are discussed according to the Euler-Bernoulli beam theory. The 'Euler-Bernoulli' or 'simple' beam theory neglects the effect of rotational inertia and shear deformation and is valid for beams with a large length-to-height ratio and sufficiently small bending [39, p.207]. If a more general beam model is required that also accounts for rotational inertia and shear deformation, the Timoschenko beam theory can be employed, which contains the simple Euler-Bernoulli theory as a special case [40, p.287]. The derivation of the Euler-Bernoulli beam equations from bending moment and shear force considerations, as well as the solutions for standard boundary conditions such as a one side clamped cantilever beam may be found in many textbooks on technical mechanics [41, 39, 40]. The general differential equation for a Euler-Bernoulli beam is given by [39, p.208]:

$$\frac{\partial^2}{\partial x^2} \left((EI)(x) \frac{\partial^2 u(x,t)}{\partial x^2} \right) + \rho(x) S(x) \frac{\partial^2 u(x,t)}{\partial t^2} = F(x,t) \tag{2.13}$$

with $u(x,t)$ being the vertical displacement of the neutral axis of the beam at position x and $F(x,t)$ a vertical force per unit length (see Figure 2.6). In this context the neutral axis is the axis in the center of the beam that is neither compressed nor dilated [25, p.4] when the beam is bent. If the beam is uniform, meaning that the cross section area S, mass density ρ, and flexural rigidity EI, i.e. the product of Young's modulus E and the second moment of area of the cross section I are constant along its length, this reduces to

$$EI\frac{\partial^4 u(x,t)}{\partial x^4} + \rho S\frac{\partial^2 u(x,t)}{\partial t^2} = F(x,t). \tag{2.14}$$

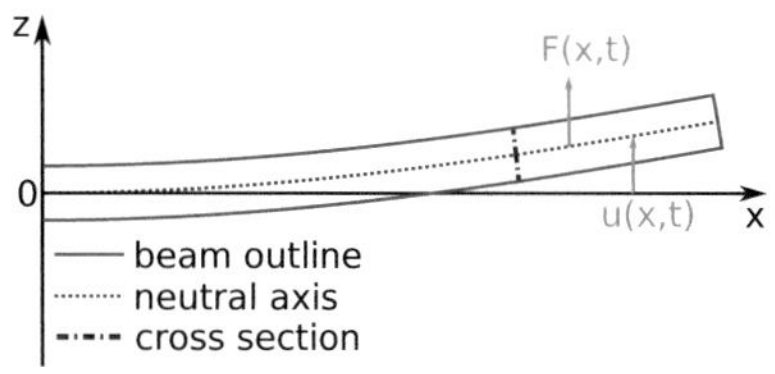

Figure 2.6: Sketch of a bend cantilever beam to illustrate the coordinates and directions used in the simple beam model. This model assumes that any cross section of the beam oriented orthogonal to the neutral axis in the equilibrium state remains orthogonally orientated with respect to the neutral axis during bending.

2.3.1 Dynamic free solution

To derive the dynamic solutions for a beam with one fixed and one free end, the above differential equation 2.14 has to be solved with the appropriate boundary conditions. The fixed end can neither be subject to deflection nor rotation, so both displacement and slope at this end are equal to zero. In contrast, the free end can move freely and neither bending moment nor vertical shear force are present. This results in the boundary conditions [39, p.209]:

$$u(x,t)|_{x=0} = 0; \quad \left.\frac{\partial u(x,t)}{\partial x}\right|_{x=0} = 0; \quad \left.EI\frac{\partial^2 u(x,t)}{\partial x^2}\right|_{x=L} = 0; \quad \left.EI\frac{\partial^3 u(x,t)}{\partial x^3}\right|_{x=L} = 0. \tag{2.15}$$

Using a harmonic separation ansatz

$$u(x,t) = u_n(x)cos(\omega_{0,n}t + \phi) \tag{2.16}$$

as well as the condition for a free system $F(x,t) = 0$ and the definition

$$\beta_n^4 = \frac{\omega_{0,n}^2 \rho S}{EI} \tag{2.17}$$

together with equation 2.14 this results in the time independent differential equation:

$$\frac{\partial^4 u_n(x)}{\partial x^4} - \beta_n^4 u_n(x) = 0. \tag{2.18}$$

A solution to this equation may be derived with an exponential ansatz [40, p.283] and can be written as [25, p.10]:

$$\begin{aligned} u_n(x) = & C_1\left(cos(\beta_n x) + cosh(\beta_n x)\right) + C_2\left(cos(\beta_n x) - cosh(\beta_n x)\right) + \\ & C_3\left(sin(\beta_n x) + sinh(\beta_n x)\right) + C_4\left(sin(\beta_n x) - sinh(\beta_n x)\right) \end{aligned} \tag{2.19}$$

with constants C_1 to C_4. Using the afore mentioned boundary conditions results in

$$C_1 = C_3 = 0; \quad C_4 = \frac{sin(\beta_n L) - sinh(\beta_n L)}{cos(\beta_n L) + cosh(\beta_n L)} C_2 \tag{2.20}$$

and the characteristic implicit equation

$$cos(\beta_n L) cosh(\beta_n L) = -1. \tag{2.21}$$

This equation yields an infinite number of solutions. The solutions can be approximated by numerical methods and the first three are found to be [25, p.11]:

$$\beta_1 L \approx 1.875; \quad \beta_2 L \approx 4.694; \quad \beta_3 L \approx 7.855; \quad ... \tag{2.22}$$

Considering all the above the solutions for the shape of mode $u_n(x)$ and, since it is required later on, its first derivative $u'_n(x)$ are:

$$\begin{aligned} u_n(x) = & C_2[cos(\beta_n x) - cosh(\beta_n x) + \\ & \frac{sin(\beta_n L) - sinh(\beta_n L)}{cos(\beta_n L) + cosh(\beta_n L)}(sin(\beta_n x) - sinh(\beta_n x))] \end{aligned} \tag{2.23}$$

$$\begin{aligned} u'_n(x) = & -C_2 \beta_n[sin(\beta_n x) + sinh(\beta_n x) + \\ & \frac{sin(\beta_n L) - sinh(\beta_n L)}{cos(\beta_n L) + cosh(\beta_n L)}(cosh(\beta_n x) - cos(\beta_n x))]. \end{aligned} \tag{2.24}$$

The remaining constant C_2 can be used to fit these solutions to the absolute amplitudes required by experimental data. Furthermore, the eigenfrequencies

$$\omega_{0,n} = \beta_n^2 \sqrt{\frac{EI}{\rho S}} \tag{2.25}$$

can be obtained by rearranging equation 2.17.

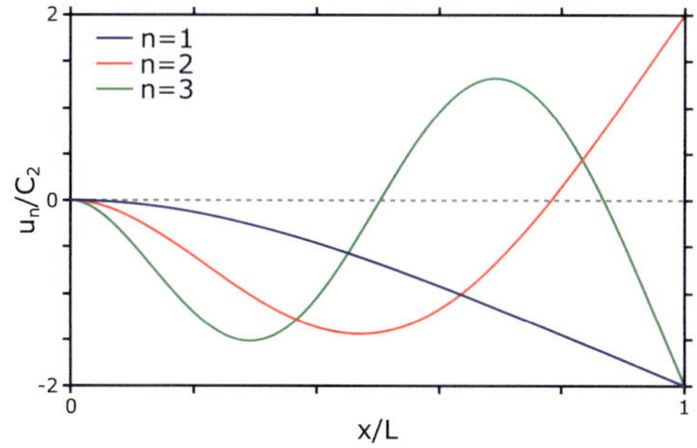

Figure 2.7: Mode shapes $u_n(x)$ for the first three flexural modes of a cantilever beam as given by equation 2.23.

Here a fundamental difference between this continuous model and the simple harmonic oscillator is obvious: the existence of not only one but an infinite number of eigenfrequencies and corresponding flexural oscillation modes. The mode shape is different for each mode as shown in figure 2.7 and the n-th mode exhibits $n-1$ nodal points. The simple beam theory essentialy treats the original three dimensional problem in a projection to two dimensions. A real three-dimensional beam will show additional eigenfrequencies and corresponding modes of oscillation. These include torsional and lateral oscillations as shown in figure 2.8 but also axial and even more complicated combined oscillation modes [42].

2.3.2 Static solution for a point-like force

If the static solution $u(x)$ for a static point-like vertical force acting on the free end of a cantilever $F(x,t) = F_0\delta(x-L)$ is derived, the time derivatives are equal to zero and the differential equation 2.14 then reads

$$\frac{\partial^4 u(x)}{\partial x^4} = \frac{F_0}{EI}\delta(x-L). \tag{2.26}$$

Although this differential equation may be formally solved by integration using singularity functions [43, pp.607-610], it is possible to choose a simpler approach for this problem: to account for the external force by using a corresponding boundary condition and then solve the homogeneous equation

$$\frac{\partial^4 u(x)}{\partial x^4} = 0 \tag{2.27}$$

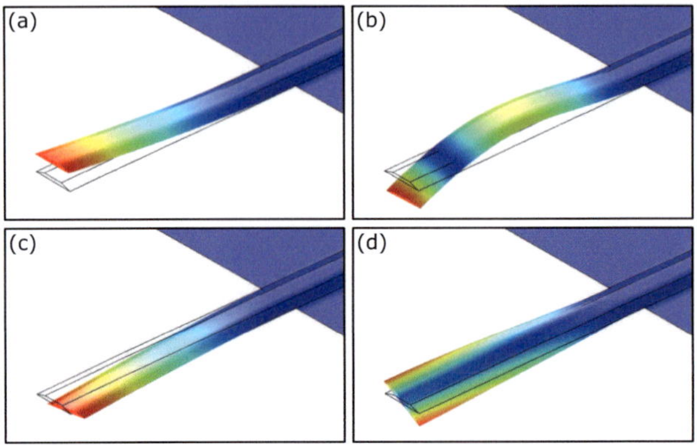

Figure 2.8: The first oscillation modes of a three dimensional cantilever beam: (a) and (b) correspond to the first and the second flexural, (c) to the first lateral and (d) to the first torsional mode respectively. The simulations were based on a 215 µm long, 3 µm high Si cantilever beam with an isosceles trapezoid cross section (upper length 20 µm and lower length 40 µm). The finite element method simulation software 'COMSOL Multiphysics' was used for the simulation. The colour corresponds to the deflection from the equilibrium position and the images are ordered according to the respective eigenfrequencies with (a) being the lowest and (d) the highest.

by simple integration [44, p.149]. The integration constants C_1 to C_4 of the general solution

$$u(x) = \frac{1}{6}C_1x^3 + \frac{1}{2}C_2x^2 + C_3x + C_4 \tag{2.28}$$

are again determined by boundary conditions. In equilibrium the external vertical force acting on the free end of the cantilever is compensated by an opposing vertical shear force of equal magnitude. The boundary conditions then read [44, p.149]:

$$u(x)\big|_{x=0} = 0; \quad \left.\frac{\partial u(x)}{\partial x}\right|_{x=0} = 0; \quad EI\left.\frac{\partial^2 u(x)}{\partial x^2}\right|_{x=L} = 0; \quad EI\left.\frac{\partial^3 u(x)}{\partial x^3}\right|_{x=L} = -F_0. \tag{2.29}$$

The solution can the be derived as [44, p.149]

$$u^{stat}(x) = \frac{F_0}{6EI}\left(3Lx^2 - x^3\right). \tag{2.30}$$

The static spring constant k^{stat} is defined according to Hooke's law as the proportionality factor between the deflection u and the applied force F_0

$$F_0 = k^{stat} \cdot u. \tag{2.31}$$

Hence, the static spring constant of a cantilever for a vertical force acting on its free end $x = L$ is given by

$$k_L^{stat} = 3\frac{EI}{L^3}. \tag{2.32}$$

2.3.3 Transfer to harmonic oscillator model

In order to use the results of the Euler-Bernoulli beam model in the harmonic oscillator model, one could simply use the static spring constant k^{stat} given by equation 2.32. It is found, however, that this static spring constant does not correctly describe the dynamic behavior of the system even for the fundamental mode, which is used in standard dSFM. The reason for this are the different shapes of the static and dynamic beam deflection and the related different potential energies. A mathematical description of this difference is given by the inverse static spring constant being equal to the sum of all inverse dynamic spring constants [45]. Numerical calculations reveal, however, a simple relation between the static k_L^{stat} and the dynamic spring constant $k_{1,L}^{dyn}$ of the fundamental mode for a cantilever with a constant cross section area along its length with the end of the cantilever as point of reference [45]:

$$k_{1,L}^{dyn} \approx 1.03 k_L^{stat}. \tag{2.33}$$

Furthermore, the effective mass that can be employed in the harmonic oscillator model is not given by the total mass of the cantilever as not the whole mass of the cantilever is accelerated equally during the oscillation cycle. By combining the equation 2.5 for the eigenfrequency of the harmonic oscillator with equations 2.22, 2.25, 2.32 and 2.33 and solving for the effective mass m_{eff} of the harmonic oscillator, the resulting relation to the real mass $M = \rho S L$ of the cantilever beam is given by:

$$m_{eff} = \frac{k_{1,L}^{dyn}}{\omega_{0,n}^2} = 1.03\frac{3\rho SL}{(\beta_1 L)^4} \approx 0.25M \tag{2.34}$$

Up to now the effect of the tip mass was neglected. To take the tip into account, we can approximate a tip at the end of the cantilever beam by adding its total mass to the effective mass [46, 47]. If the tip is very heavy or very large, however, it will have a non-negligible influence on the oscillatory properties and modal shape of the cantilever especially for higher modes and more sophisticated calculation methods have to be used [48]. Since the tips that were used for the experiments presented in this thesis are comparatively small and light, their influence was neglected in the corresponding calculations.

If higher modes of the cantilever are considered, the behavior of each mode can be approximated by a separate harmonic oscillator with a corresponding effective dynamic spring constant [24]. The effective dynamic spring constant for the second flexural mode $k_{2,L}^{dyn}$ referring to its free end can again be numerically related to the static one [45]:

$$k_{2,L}^{dyn} \approx 40.2 k_L^{stat} \tag{2.35}$$

Finally it should be mentioned that in the above considerations neither damping nor mechanical excitation of the sensor is considered. In principle, both internal and external damping [49, 50] as well as an external excitation of the cantilever beam can be included in the beam theory. This leads to an increased complexity of the involved mathematics [51], however, and the more simple free beam model was already found to yield satisfactory results [47, 45] in the case of very low damping and the use of more complex models is often unnecessary.

2.4 Spring constant calibration

A quantitative evaluation of frequency shift data acquired with a dSFM sensor is based on the knowledge of its spring constant k according to equation 2.11. Even if this simple formula cannot be used, the more complex evaluation methods also require this important quantity [9]. Since the harmonic oscillator model itself does not directly provide any means to determine the spring constant k and the effective mass m_{eff} of the system independently, one has to acquire the spring constant by different methods. There are various methods available to derive the spring constant based on geometry [52], thermal noise [53, 54], viscous damping in air [55], added masses [46], reference cantilevers [56] or the sensor's mass and resonance frequency [57]. The choice of the method depends on experimental conditions and available measurement techniques. Since in this work mostly modified sensors were employed, their geometric and dynamic properties could be measured in the course of the sensor fabrication. Therefore, a combination of these properties was chosen to determine the static spring constant.

A starting point for the calculation of static spring constants is given by equation 2.32. However, the use of this expression has several practical disadvantages: The Young's modulus E depends on the crystal orientation, which is usually unknown for commercially available cantilevers. Furthermore, the second moment of area I is proportional to the cube of the height h^3, a quantity which can be measured only with a comparatively high error.

In order to avoid these sources of uncertainty, the measured resonance frequency of the first flexural mode of the cantilever $\omega_{0,1}^* \approx \omega_{0,1}$ can be used. Rewriting equation 2.25 yields

$$EI = \frac{\omega_{0,1}^2 \rho S}{\beta_1^4}. \tag{2.36}$$

Combining equation 2.36 with equation 2.32 results in the expression

$$k_L^{stat} = \frac{3\omega_{0,1}^2 \rho S L}{(\beta_1 L)^4} \approx \frac{3\omega_{0,1}^2 \rho S L}{(1.875)^4} \tag{2.37}$$

for the desired static spring constant, making the knowledge of Young's modulus E unnecessary. Instead, it includes the mass density ρ, which is readily available, as the cantilever material is usually known. Furthermore, this equation depends only linearly on all spatial dimensions of the cantilever thus reducing the influence of their measurement errors. As this result is also based on the mass $M = \rho S L$ and the resonance frequency of the sensor, it is similar to the last method mentioned above [57]. Equation 2.37 is valid for uncoated cantilevers. If cantilevers with a reflective coating are used, this has to be taken into account [57].

It has to be noted that the simple beam model used to derive these equations assumes a point-like force acting on the free end of the cantilever. To be more precise this point-like force in the effectively two-dimensional simple beam model is rather a line-like distributed force acting on whole width of the cantilever's cross section. Additionally, a real dSFM tip always has a certain spatial extent so it already distributes the force across the area of its connection to the cantilever. However, in dSFM usually concentrated forces acting on the apex of a dSFM sensor's tip are considered. This simplification to regard forces acting on the tip of a dSFM as a distributed force is commonly accepted in literature and generally yields good results [25, 58, 57, 59, 52, 45] and is also used in this work. Furthermore, the tip of common dSFM sensors is usually positioned along the vertical symmetry axis of the cross section on the bottom of the cantilever. Therefore, an applied force does not result in an torque about this axis, thus preventing torsional deformations of the beam.

2.5 Magnetic force microscopy

If the interaction tip of a SFM sensor partly consists of ferromagnetic material, it is sensitive to magnetic forces and can be used to map the magnetic stray fields close to the surface of samples. This special type of SFM is called magnetic force microscopy (MFM) and first implementations of static [60] and dynamic mode [33] MFM have already been presented shortly after the publication of Binnig's first atomic force microscopy concept. From these beginnings, MFM has become a valuable tool to study various magnetic systems, for example magnetic domain walls [61], vortices on superconductors [62] and, important for industrial applications, magnetic recording media [38]. Recently, MFM has been reported to be able to support the study of skyrmion lattices in chiral magnets [63].

2.5.1 Magnetic tip-sample interaction

In MFM the desired signal is created by the magnetostatic interaction between the magnetic stray field above the surface of a sample and the ferromagnetic material of the probe tip. The magnetostatic scalar potential Φ_s of a ferromagnetic sample with magnetization vector field $\vec{M}_s$ can be derived from Maxwell's equations and may be written as [64, p.230]:

$$\Phi_s(\vec{r}) = \frac{1}{4\pi}\left(\int_{S_s}\frac{\vec{s}'\cdot\vec{M}_s(\vec{r}')}{|\vec{r}-\vec{r}'|}d^2\vec{r}' - \int_{V_s}\frac{\vec{\nabla}'\cdot\vec{M}_s(\vec{r}')}{|\vec{r}-\vec{r}'|}d^3\vec{r}'\right) \tag{2.38}$$

with $\vec{s}'$ being the outward-pointing normal unit vector of the sample surface and the integrations cover the whole surface S_s and volume V_s of the sample respectively. Defining the magnetic volume charge density $\rho(\vec{r}) = -\vec{\nabla}\cdot\vec{M}_s(\vec{r})$ and magnetic surface charge density $\sigma(\vec{r}) = \vec{s}\cdot\vec{M}_s(\vec{r})$ in analogy to electrostatic fields [8, p.104] gives a descriptive interpretation of the two integrals. Surface charges arise from magnetization components perpendicular to the sample surface and their contribution to the magnetostatic potential is given by the first two-dimensional integral, whereas the volume charges are created by internal divergences of the sample magnetization and are covered by the three-dimensional integral [30]. The sample's magnetic stray field $\vec{H}_s$ is then simply the negative gradient of this magnetostatic potential [30]:

$$\vec{H}_s(\vec{r}) = -\vec{\nabla}\Phi_s(\vec{r})\,. \tag{2.39}$$

It has to be noted that different magnetization configurations can lead to the same magnetic stray field. Therefore, it is not possible to reconstruct a sample's magnetization vector field $\vec{M}_s(\vec{r})$ by magnetic force microscopy measurements alone [8, p.104].

If mutual perturbations of the magnetization of sample and tip are neglected, the magnetostatic energy Ψ of the interaction of a ferromagnetic tip with a ferromagnetic sample can be derived using the reciprocity principle [65, 66] and be written as [67, p.80]:

$$\Psi = -\mu_0\int_{V_t}\vec{M}_t(\vec{r}')\cdot\vec{H}_s(\vec{r}')d^3\vec{r}' \tag{2.40}$$

with the magnetic field $\vec{H}_s(\vec{r}')$ of the sample as given by equation 2.39. $\vec{M}_t$ denotes the magnetization vector field of the tip and the integration covers its volume V_t.

In a magnetic force microscopy experiment the position of the tip is not constant with respect to the sample. Therefore, a parametrization of the magnetostatic energy with tip position is favorable [66]:

$$\Psi\left(\vec{r}\right) = -\mu_0 \int_{V_t} \vec{M}_t\left(\vec{r}^{\dagger} + \vec{r}\right) \cdot \vec{H}_s\left(\vec{r}^{\dagger} + \vec{r}\right) d^3\vec{r}^{\dagger}. \tag{2.41}$$

Here $\vec{r}$ denotes the position of the tip's apex, $\vec{r}^{\dagger}$ is an coordinate within a coordinate system fixed to the tip with $\vec{r}$ as origin. As the magnetization vector field of the tip does not change with respect to $\vec{r}$ it can be convenient to directly recalculate it to the tip's coordinate system $\vec{M}_t\left(\vec{r}^{\dagger} + \vec{r}\right) = \vec{M}_t^{\dagger}\left(\vec{r}^{\dagger}\right)$ [66]. Equation 2.41 may be transformed into a equivalent expression by using equation 2.39 and integration by parts [67, p.80]:

$$\Psi\left(\vec{r}\right) = \mu_0 \left(\int_{S_t} \vec{t}^{\dagger} \cdot \vec{M}_t\left(\vec{r}^{\dagger} + \vec{r}\right) \Phi_s\left(\vec{r}^{\dagger} + \vec{r}\right) d^2\vec{r}^{\dagger} - \int_{V_t} \Phi_s\left(\vec{r}^{\dagger} + \vec{r}\right) \vec{\nabla} \cdot \vec{M}_t\left(\vec{r}^{\dagger} + \vec{r}\right) d^3\vec{r}^{\dagger} \right) \tag{2.42}$$

with $\vec{t}^{\dagger}$ being the outward-pointing normal unit vector of the tip surface. The integrals cover the whole surface S_t and volume V_t of the magnetic tip respectively. The surface integral describes the interaction of the magnetic stray field of the sample with the magnetic surface charges of the tip and the volume integral covers the respective interaction with the volume charges. Finally the force acting on the MFM sensor due to the interaction between tip and sample can be calculated by

$$\vec{F}_{ts}\left(\vec{r}\right) = -\vec{\nabla}\Psi\left(\vec{r}\right). \tag{2.43}$$

The above expressions can in principle be used for a quantitative evaluation of MFM data if the magnetization vector field $\vec{M}_t\left(\vec{r}\right)$ of the tip is known. Unfortunately, typical MFM tips consist of a sharp tip that is coated with ferromagnetic material and the internal crystalline and domain structure of this coating is unknown, thus rendering quantitative data evaluation impossible.

2.5.2 Point probe approximation

A way to facilitate data interpretation was introduced by U. Hartmann who proposed to not consider the whole volume of the tip, but instead only the part of the tip's magnetic material that effectively contributes to the magnetic tip-sample force in the decaying stray field of the sample. The non-contributing parts of the magnetic material of the tip are neglected. He proposed further to approximate the whole magnetic response of a magnetic tip by modeling this magnetically active volume of the tip by a homogeneously magnetized prolate spheriod particle of suitable dimensions at the end of an otherwise nonmagnetic tip [68]. He also proposed a further simplification for the case that certain experimental preconditions with respect to the tip and stray field geometry are fulfilled: the point probe approximation [69]. This approximation can be derived from a multipole expansion of the magnetostatic energy. The resulting effective monopole and dipole moments are projected into a virtual infinitesimal small magnetic probe located at a certain distance from the sample to approximate the effective magnetic moment of the part of the tip material that contributes to the magnetic signal in MFM [30]. The force acting on such a probe when subjected to a magnetic field can be described by [68, 70, 71]:

$$\vec{F}_{ts}\left(\vec{r}\right) = -\mu_0(q - \vec{m}\cdot\vec{\nabla})\vec{H}\left(\vec{r}\right) \tag{2.44}$$

where q and $\vec{m}$ are the effective monopole and dipole moments of the tip, respectively. The parameters of this approximation, the effective magnetic monopole and dipole moments and the point probe-sample separation, can be determined by fits to experimental data on well defined stray fields [71, 72]. It has been shown that either the monopole or the dipole description is sufficient to describe a pyramidal magnetic tip [71]. Assuming a tip with pure effective monopole moment q, and a tip oscillation along one of the basic directions $s = x, y, z$ the frequency shift in MFM experiments with such a tip can be calculated by combining equation 2.44 with the equations 2.11 and 2.12 for the frequency shift in dSFM resulting in:

$$\Delta f = \frac{qf_0}{2k}\mu_0\frac{\partial H_s}{\partial s}. \tag{2.45}$$

An image of a commercially available MFM tip[1] and a visualization of both the effective magnetic volume and the point probe approximation can be found in figure 2.9.

Please note that the magnetically active volume of the tip is usually dependent on the actual stray field geometry of the sample in which case the parameters of the point probe approximation depend on this stray field geometry as well.

2.5.3 Separation of forces

As already mentioned before, one of the challenges when trying to get a purely magnetic signal with a dSFM sensor is the fact that typically several different forces act on the

[1]HR-MFM 225C0.7-ML3-R by Team Nanotec GmbH

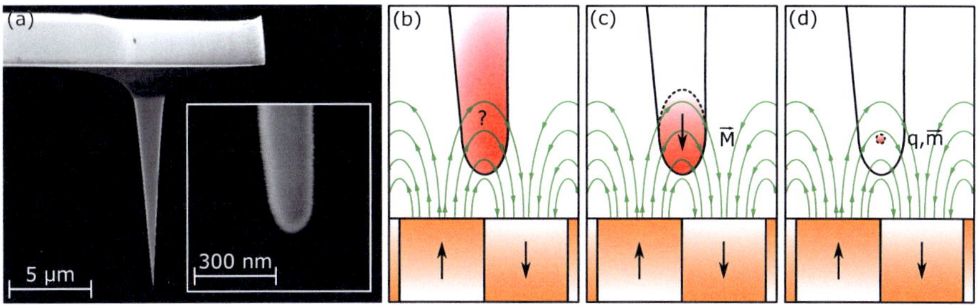

Figure 2.9: Simplifications to approximate a magnetic tip's interaction with a magnetic stray field: (a) scanning electron microscopy image of a real MFM tip; the inset shows the tip's apex with higher magnification. (b) Sketch of a tip with an unknown magnetization structure. (c) First approximation: prolate spheriod particle with homogenous magnetization $\vec{M}$; the remainder of the tip being magnetically ineffective. (d) Second approximation: point probe with effective magnetic monopole moment q and magnetic dipole moment $\vec{m}$ (according to [70])

sensor simultaneously. Since magnetic interactions are comparatively weak but of long range, the typical way to reduce the number of forces acting on the tip is to measure at a certain distance from the sample surface. At large tip-sample distances only forces with long range components have to be considered. In addition to the desired magnetic forces, these long-range forces are the ever present van-der-Waals forces and electrostatic forces due to accumulated charges, polar materials or, in the case of conductors, different work functions of tip and sample.

In most cases, a removal of the contributions of the electrostatic forces is possible if both sample and tip surface are conductive and of homogenous material composition. If this is not the case a very thin surface layer of conductive material can be added. The tip and sample arrangement can then be considered to be a capacitor with a distance-dependent capacitance z with the electrostatic force given by:

$$F_{ts}^{el} = \frac{1}{2}\frac{\partial C}{\partial z}(U_{bias} - U_{contact})^2. \tag{2.46}$$

This electrostatic force has a quadratic dependence on the difference of an applied external bias voltage U_{bias} and the contact potential $U_{contact}$ originating from different work functions of tip and sample [8, p.54]. Assuming an oscillation of the dSFM sensor along the z-direction perpendicular to the sample surface and a non-vanishing, only slowly changing second derivative of the capacitance $\partial^2 C/\partial z^2$, the frequency shift induced by this force $\Delta f \propto (U_{bias} - U_{contact})^2$ still retains the quadratic dependence. Therefore, measuring the frequency shift Δf at a fixed position over the sample surface while sweeping the bias voltage will result in a parabolic Δf-U_{bias}-curve. At the vertex of this parabola, which can be determined by a parabolic fit, the bias voltage compensates the contact potential and the electrostatic force will vanish. By using this bias voltage for MFM measurements on a sample with homogenous contact potential, the influence of the electrostatic force can be minimized.

A different matter entirely is the van-der-Waals contribution, which contains information about the topography of the sample. There is no simple way to eliminate its contribution entirely, but several approaches can be used to minimize its influence. For sufficiently flat samples with homogenous surface material composition a measurement of the magnetic forces can be obtained by simply scanning at a constant height parallel to the sample surface. Neglecting atomic resolution contributions, the van-der-Waals forces should then be constant over the whole scan area. Depending on the average tip-sample separation this constant offset is either negligibly small [8, p.123] and can be ignored, otherwise it can be accounted for either by theoretical calculations [8, p.53] or measurements on sample positions without magnetic signal. Furthermore, in case of samples with a periodic magnetic structure with equal distribution of positive and negative magnetic force gradients, the offset may simply be subtracted from the data. If the sample surface is not flat, the van-der-Waals contribution will not be constant and the topography will be visible in the measurement signal for small tip-sample separations.

In this case the two-pass technique, sometimes also called 'lift-mode', can be employed [8, p.122]. In this technique, first the topography of the sample is measured at very small tip-sample separations and then this data is used to retrace the topography at a predefined sensor position above the sample surface. This way a constant average tip-surface separation is achieved resulting again in an approximately constant van-der-Waals forces contribution. Especially in the case of large, non-magnetic surface features, however, there can still be cross-talk between the magnetic and the topography signal. Furthermore, this method is very slow since two scans of the sample area are required. Additionally, the equipment has to be reasonably stable to avoid drifts between the two measurements. Recently, however, methods based on the simultaneous excitation of the first and the second flexural mode of a dSFM cantilever have been proposed to solve this particular problem [73, 74].

A way to achieve a true separation between topography and magnetic signal is by measuring the same data set twice, and reversing the tip magnetization in-between. The magnetic forces will change their direction, because of this magnetization reversal while all other forces will stay the same. On the one hand a purely magnetic signal can then be obtained by simply subtracting the two data sets. On the other hand an addition gives the signal from the remainder of the, usually topography related force gradients. This simple method has high demands on the stability and positioning mechanism of the system, however, and requires magnetic tips with well defined opposing states of magnetization [8, p.124].

2.5.4 Magnetic stability

In the discussions above it has always been assumed that the tip and sample magnetization do not influence each other. However, depending on the magnetic properties of tip and sample this assumption does not always hold, especially in close proximity of tip and sample. These mutual influences can lead to many different effects both reversible and irreversible. Examples for reversible perturbations are small rotations of the magnetization of the tip or a temporary change of the domain size of a sample in the measured area due to the magnetic stray field of the tip [8, p.116]. Irreversible changes include more drastic effects as a permanent reordering of the domain configuration of a sample or a reversal of the magnetization of the tip. Irreversible perturbations of the magnetization structure of sample or tip during the measurement process need to be prevented by an appropriate choice of the magnetic properties of the tip and a sufficient tip-sample separation. Examples for reversible and reversible perturbations are shown in figure 2.10.

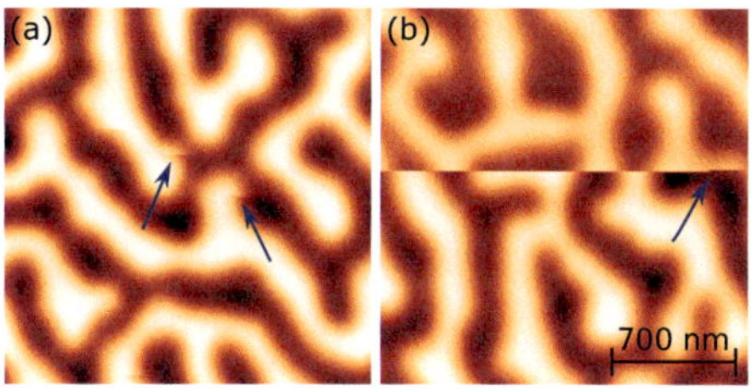

Figure 2.10: Artifacts due to mutual magnetic perturbations of tip and sample visible in MFM frequency shift data measured on a Co/Pt - multilayer sample with a commercially available MFM tip. In (a) the arrows indicate minor reversible perturbations. In (b) an irreversible perturbation is shown. At the scan position indicated by the arrow the magnetization of the tip was reversed by the magnetic stray field of the sample. This resulted in an inversion of the image contrast.

2.6 Equipment and experimental practices

After various theoretical aspects of the dSFM technique have been discussed above the experimental equipment and related to this, some of the experimental practices that are of importance for the quantitative dSFM measurements in this work are presented in the following.

2.6.1 Scanning force microscopy equipment

All of the dSFM measurements herein were conducted by using a 'hr-MFM' scanning force microscope manufactured by the company 'NanoScan AG'. This system is optimized for dynamic mode magnetic force microscopy measurements, but offers a large range of options to perform other types of dynamic and static SFM measurements under ambient pressure as well as in high vacuum. As mentioned before, the detection of the cantilever's oscillation is realized by a laser beam deflection method using a sectioned photo diode sensitive to both lateral and normal deflections. The mechanical excitation of the sensor is realized by a piezoelectric actuator. The system features three sets of feedback electronics for full frequency modulation measurement support: A phase-locked loop (PLL) can be utilized to keep a constant phase shift between periodic excitation and dSFM sensor oscillation, enabling frequency shift detection. Two other feedback circuits controll the vertical position of the dSFM sensor and the amplitude of its oscillation. The system is furthermore equipped with a linearized scanner stage with a range of 40 μm $\times$ 40 μm and a lateral resolution of up to 0.2 nm. Furthermore, a perpendicular and an in-plane magnetic field option is available. The vertical sensor position of the sensor is adjusted by a linearized piezoelectric actuator with a maximal range of 6 μm. The laser deflection system further provides the ability of a lateral movement of the laser spot in two directions to adjust its position to that of the cantilever's free end as well as a mechanism to be able to center the reflected beam on the detector. For a coarse positioning of the laser spot on the cantilever and of the cantilever with respect to the sample, the system is furthermore equipped with two optical digital cameras. The instrument's control software offers access to various experimental parameters of the system and includes additional tools like a bias sweep method to compensate for the contact potential as well as a method to acquire amplitude and phase versus driving frequency curves of dSFM sensors to characterize their dynamic behavior [75, 76].

2.6.2 Calibration of the oscillation amplitude

A first challenge for quantitative measurements with a beam deflection based dSFM instrument is the calibration of the oscillation amplitude of the cantilever. Each section of the sectioned photo diode generates a signal depending on the power of the incident laser light that reaches its surface. The position of the beam on the detector varies depending on the angle of the cantilever, so a change of the cantilever deflection will cause a displacement of the reflected laser beam on the detector's surface [77]. This displacement will change the signal generated by each detector section. By forming a difference signal of two respective halves of the detector even small cantilever deflections in the sub-Ångström range can be detected at measurement periods in the order of one millisecond [78, 79].The measurement of the cantilever's deflection with this technique is limited by thermal vibration noise, however, which causes a virtual mean deflection of $\sqrt{< s^2 >} = \sqrt{4k_BT/3k}$, with k_BT being the thermal energy and k the cantilever's spring constant [8, pp.92-95].

Now a method is needed to relate the difference signal of the detector to the actual cantilever deflection. A very simple static method consists of pressing the tip of a dSFM sensor firmly onto a hard sample surface until the cantilever bends upwards [80, 53]. For small deflections, a calibration can be derived by correlating the changing detector signal with the distance change caused by lowering the dSFM sensor. Because this method severely damages the tip, it is typically used by sacrificing the first tip of a new batch of dSFM sensors and simply assuming that the calibration will be valid for all the other sensors of this batch, as well. [80]. Since the dSFM sensors used in this work were often heavily modified this was not a viable approach for the experiments.

The dynamic calibration method that was employed in this work [80] preserves the tip and is reminiscent to an intermittent contact mode. In the absence of magnetic and electrostatic force components, and in high vacuum the interaction of the tip with a substrate is due to van-der-Waals and short range forces. The range of such forces is limited, however, and an interaction will only take place at the lowest parts of the tip's trajectory if the amplitude is much larger than the range of these forces [9]. Furthermore, theoretical considerations revealed that the frequency shift signal of dSFM sensor depends strongly on the tip-sample force at and close to the turning points of the tip's motion [9, 81]. This is the case because the velocity of the tip around these points is very low or even zero at the turning point and therefore the time for an interaction between tip and sample is increased [32]. For large amplitudes much longer than the range of these forces the upper turning point is far away from the sample and therefore the forces acting on the tip at this point are negligibly small. Therefore, the main contribution to the frequency shift results from the interactions very close to the lower turning point of the tip motion. Hence, if a dSFM sensor is operated at very large oscillation amplitudes and is approached to the sample until a frequency shift signal can be detected, the mean tip-sample separation is approximately equal to the oscillation amplitude in case no long range interactions are present.

This was utilized to implement a calibration procedure for the oscillation amplitudes of the sensors employed in this work: after the determination of the resonance frequency of a dSFM sensor, the amplitude feedback loop was used to control the excitation in order to steady the detector amplitude signal at a given set-point. Next, the phase locked loop was utilized to enable frequency shift detection and the sensor was approached to a hard, flat, homogenous and conductive surface, usually a doped silicon wafer. After the compensation of the contact potential by a bias sweep as discussed above, the sensor was moved closer to the sample until a small negative frequency shift signal was obtained. Reducing the amplitude at a fixed height results in the turning point of the tip's motion moving farther away from the sample surface, thus reducing the tip-sample interaction and the corresponding absolute value of the frequency shift signal. Therefore, the vertical piezoelectric actuator feedback loop that controls the dSMF sensor's height was set up to keep the frequency shift at a fixed small negative value, so that it responded to a decreasing amplitude set-point by approaching the sensor to the sample to keep the interaction constant. By sweeping the detector amplitude signal set-point while recording the position of the vertical piezoelectric actuator needed to obtain a constant frequency shift, data sets as exemplarily shown in figure 2.11 were obtained. For large oscillation amplitudes the change in the amplitude is approximately equal to the change in the mean tip-sample separation. This results in a linear relationship between the vertical sensor position and an uncalibrated amplitude detector signal and the amplitude calibrations for the sensors were based on the required equality.

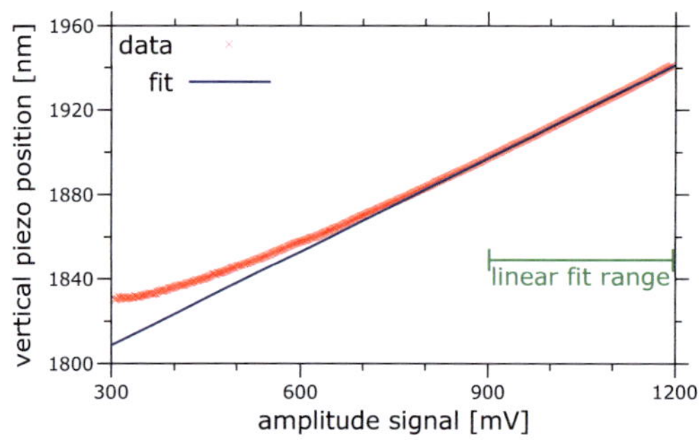

Figure 2.11: Exemplary data for the dynamic amplitude calibration of a dSFM sensor. The fit range was chosen by hand to correspond to the upper linear part of the data. The slope of the linear fit corresponds to the calibration factor, in this case a value of 145.1 nm/V was found.

In figure 2.11 at large amplitudes the approximately linear relationship can be recognized. In contrast, at low amplitudes the behavior deviates visibly. The calibration factor to relate the detector amplitude signal to the oscillation amplitude is obtained as the slope of a linear fit to the upper part of the data. The usually unknown exact relationship between the tip-sample interaction and the momentary tip-sample distance results in a difficulty to quantify error of the thus obtained calibration. This trade-off between precision, tip preservation and practicability is necessary, however. Please note that the vertical sensor position and therefore this amplitude calibration is also subject to uncertainties of the piezoelectric actuator positioning system. These will be discussed in more detail below.

Of course, the calibration factor is only valid if the position of the laser beam on the cantilever is the same in each measurement. This is no trivial task, however, since the precise position of the laser spot on the cantilever is not easily visible in the optical cameras of the dSFM equipment and only the overall intensity of the beam on the detector and the mean position of the laser spot on the detector are available information.

For maximum reproducibility the laser beam was first positioned on the cantilever until the detector shows the maximum light intensity. Then, the beam was moved downward which corresponded to the direction to the cantilevers free end, until the intensity of the reflected laser light on the detector began to fade. Next, the laser spot was moved left and right to find the maximum intensity again, which corresponds to a position on the middle of the cantilever. These steps were repeated until the end of the cantilever was found, signified by an intensity drop when the laser spot was moved in any other direction than upwards towards the base of the cantilever. Finally, the detector system was readjusted so that the mean position of the beam was at the center of the detector.

2.6.3 Mean tip-sample separation

Another experimental value of interest for quantitative measurements is the mean position of the sensor's tip above the sample surface. A relative separation of different vertical sensor positions is easy to obtain, since the equipment's vertical piezoelectric actuator constitutes a precise positioning mechanism. The mean tip-sample separation is challenging to determine experimentally, however, since instabilities and the influence of various force types make the determination of a 'zero-distance' difficult.

For a derivation of the tip-sample separation, first a theoretical definition of the location of the 'surface' or 'zero distance' of a sample in a SFM context has to be given. This problem can be approached by revisiting the model already discussed in section 2.1, where the tip-sample force is derived from a Lennard-Jones potential with a long-range van-der-Waals volume contribution [9] while neglecting additional long range magnetic or electrostatic forces. The 'surface' or 'zero distance' of a homogenous sample in the absence of additional forces can then be defined as a characteristic point of this force versus sensor position curve, such as its minimum [77].

A determination of such a force versus sensor position curve is complicated due to the existence of instabilities in the measurement process. In static SFM the sensor's deflection is proportional to the tip-sample force, with the spring constant of the sensor as proportionality constant. Therefore, using this method to directly measure the force-distance curve seems like a viable approach. As the sensor approaches the sample, however, the gradient of the attractive tip-sample force will increase. If it becomes larger than the spring constant of the SFM sensor, the tip will jump into contact with the sample. When retracting the sensor again the tip will adhere to the sample surface until the restoring force exerted by the SFM sensor's cantilever beam is higher than the attractive tip-sample force, at which point the tip will detach and jump to a new equilibrium position [8, p.63].

This instability is commonly referred to as 'jump-to-contact' or 'snap-in' and results in a normal deflection versus vertical sensor position curve of the SFM sensor similar to the left hand side of figure 2.12. If the static spring constant of the sensor is known, such a curve can be recalculated into a force versus sensor position curve by simple multiplication.

The 'jump-to-contact' instability disrupts the measurement of a force versus sensor position curve making it difficult to determine the real distance dependance of the tip-sample force from such measurements. This prevents a determination of the 'zero-distance' as a characteristic feature of such a force-position curve [77]. To increase the stability of the measurement, cantilevers with higher spring constants can be used, however the resulting lower deflection signal is also harder to detect.

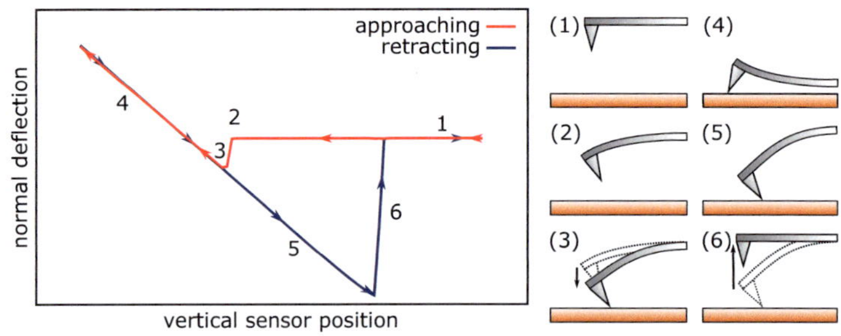

Figure 2.12: Static normal deflection versus vertical sensor position curve of a scanning force microscopy sensor approaching a neutral sample surface. This illustrates the 'jump-to-contact' instability in static mode SFM. The curve can be divided in multiple sections: (1) equilibrium position, (2) attractive tip-sample force starts to bend the cantilever, (3) 'jump-to-contact', (4) repulsive tip-sample force, (5) adhesion and (6) detachment and return to equilibrium position. Please note that the deflection corresponding to (2) is tiny and that the deformation of the cantilever is exaggerated for better visualization in the sketches on the right hand side.

The 'jump-to-contact' instability also exists in the case of dynamic mode measurements. Here, in addition to the above mentioned criterion for its occurrence, the restoring force exerted by the cantilever at the lowest point of the tip's trajectory has to be lower than the attractive tip-sample force [9]. The effect in this case is even worse since an adhesion of the tip to the sample will hamper the oscillation of the sensor, thereby preventing further dSFM measurements. As the restoring force is dependent on both the oscillation amplitude and the spring constant of the cantilever, a way to prevent this instability is given by employing high oscillation amplitudes or stiffer cantilevers. It has to be kept in mind, however, that stiffer cantilevers also reduce the frequency shift signal and if the employed oscillation amplitudes are in the same order as the length scale for changes of the measured interaction, the quantitative evaluation of the experimental data is more complicated.

Besides the 'jump-to-contact' instability, the derivation of force values from dynamic mode SFM measurements requires complex methods if short range forces are involved as mentioned before in section 2.2.3 and are therefore not practical for a fast determination of the tip-sample distance. Instead it would be possible to use characteristic features of corresponding frequency shift signal - position curve. With the sensors and operation parameters that were used in this work, however, corresponding measurements are interrupted by a jump-to-contact before any characteristic features are visible in the frequency shift signal curves. Additionally, the influence of long range magnetostatic force contributions, which can be either positive or negative depending on the magnetic properties of tip and sample, further complicates such measurements. Therefore, another method had to be found for a practicable mean tip-sample separation measurement, especially if long range magnetic forces are present.

As discussed in section 2.2.3 an additional measurement signal available in frequency modulation dSFM is the excitation amplitude needed to achieve a defined oscillation amplitude of the sensor which can be used to measure energy dissipation. In a vacuum environment this dissipation consists of a constant contribution from the internal damping of the cantilever and possible contributions from tip-sample interactions. Such dissipative tip-sample interactions have been found to play a role at close proximity to the sample [82] and can be used as an independent signal for the determination of the tip-sample separation [8, p.86]. Any dissipation signals observed in the experiments within this thesis before an interruption by a jump-to-contact were found to depend heavily on the settings of the PLL of the dSFM equipment, however, and even vanished for an appropriately adjusted PLL. They were therefore considered as artifacts of the measurement and not examined further.

None of the above discussed 'common' methods to determine the mean tip-sample distance were found to be viable options for the experiments in this work. Hence, a different method to derive the vertical position of the sensor above the sample was chosen: the occurrence of the 'jump-to-contact' instability itself was taken as a point of reference. The zero-distance is then defined by the vertical sensor position corresponding to the jump-to-contact minus the oscillation amplitude. The oscillation amplitude is subtracted since the point of reference for the second condition of the jump-to-contact to occur is the lowest point of the tip's trajectory and not the mean position.

It is emphasized that this method is not recommended for standard dSFM tips as in each measurement the tip touches the sample surface and can get damaged. Due to the improved durability of the special tips used in this work, however, this was a viable method.

The uncertainties of this method are difficult to determine. A first possible source of errors is the vertical piezoelectric actuator itself. Piezoelectric actuators are subject to piezoelectric creep, non-linearities, hysteresis and thermal drift of the whole setup [8, p.9]. Since the scanner is linearized [75] and was never operated close to its maximal movement range the influence of the first three of these sources of errors should be negligible. The influence of the thermal drift and a corresponding relative error on the height measurements is extremely difficult to quantify and additionally, it depends on the temperature of the environment. Therefore, this small relative error was neglected for the error considerations in this work.

From experimental observations, the margins of error of this method from other sources were roughly estimated to be ± 5 nm for the magnetic force microscopy measurements in this work. This includes a random error due to the measurement process as well as systematic off-sets, for example the unknown surface roughness of the sample and the different influence of distance-dependent magnetic forces at various positions of the sample and an uncertainty in the amplitude calibration.

Furthermore, this method to determine a zero-distance is somewhat arbitrary with respect to a zero-distance determined from a real distance dependence of the tip-sample force as discussed above and the result depends on both the properties of the cantilever and the operation parameters of the setup. Therefore, it cannot be considered a true determination of the mean tip-sample separation. It offers a reproducible way to position a sensor above the surface of a magnetic sample, however, which was needed for most experiments shown in this thesis. By using identical operation parameters for each position determination, this method provided good results in this aspect and, within the estimated margins of error, it even provided reasonable results for the mean tip-sample separation as is demonstrated later in section 5.5.6.

3 Sample preparation and measurement techniques

In the previous chapter the dSFM equipment that was used to conduct the dSFM experiments has already been presented. In addition to the ability to examine force-related properties of samples, the dSFM equipment offers many possibilities to characterize the equipped dSFM sensors as well. Further sample preparation and measurement techniques were needed for the experiments in this thesis, however, to fabricate novel dSFM sensor types and other samples based on commercially available cantilevers and iron-filled carbon nanotubes. This includes various scanning electron microscopy and focused ion beam methods as well as micromanipulation operations. In addition to the already mentioned dSFM methods, two other standard methods to characterize samples were employed: the geometrical measurement capabilities of a scanning electron microscope and a special sample stage to conduct vibration measurements to gather information on the dynamical properties of cantilevers and more complex specimens.

3.1 Scanning electron microscopy

Scanning electron microscopy (SEM) uses electrons to create images of conducting samples in a vacuum environment with a spatial resolution down to a few nanometers and magnification factors of 300000x and even more [83, p.63]. In SEM electrons are generated by an electron source, then accelerated and focused into a primary beam by an electromagnetic lens system. This primary beam is then scanned across the surface of a sample and images are created by relating the momentary beam position to signals from detectors, which gather electrons emitted from the sample surface. As charging effects can drastically disturb the imaging process, the samples studied with SEM should be either conductive or coated with an thin layer of conductive material.

When the primary beam electrons (PE) meet the sample surface, their high kinetic energy between 1 to 30 keV leads to a certain penetration depth into the material. On their way through the sample, the PE are subject to elastic and inelastic scattering processes that

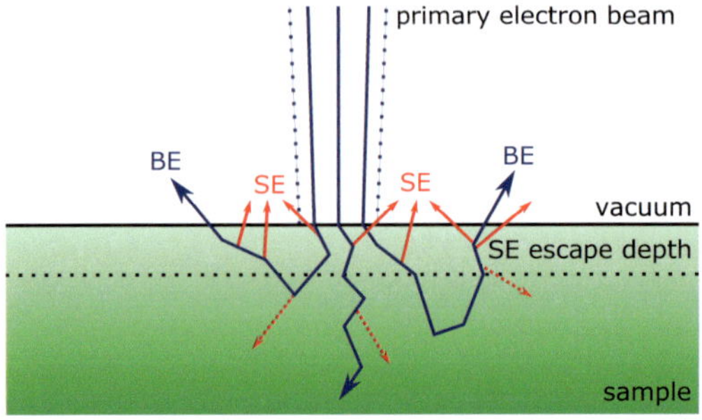

Figure 3.1: The creation of secondary electrons (SE) from primary beam electrons: Secondary electrons are generated along the whole path of the primary beam electrons through the sample material, but only those close to the surface may escape the sample and contribute to the SE signal. All others are reabsorbed. Primary beam electrons that leave the sample surface due to scattering in the sample material are called back-scattered electrons (BE) (after [83]).

change their trajectory. In case of inelastic scattering the electrons are also decelerated. Whereas some of the PE lose all their kinetic energy and stay in the material, others change their movement direction enough to leave the sample surface again after traveling through the sample. These electrons are then called back-scattered electrons (BE) and they can emerge from the surface at some lateral distance away from the primary beam. The maximal volume in which it is possible for PE to move and interact inside the sample until they are either absorbed or leave the sample surface is called interaction volume [83, pp.9-15].

Secondary electrons (SE) originate from inelastic scattering processes of the PE inside the sample and have energies lower than 50 eV. Because of their low energy they can only leave the sample if they are generated close to its surface. SE that escape from the sample surface are subsequently accelerated towards a detector by a positively biased electrode and the detector signal from the SE and incident BE is used to generate the image contrast [84]. Please note that there are also other kinds of detectors that are optimized for BE detection, but in this work mainly SE contrast was used.

Because of the short escape depth of the SE electrons, those that contribute to the SEM signal are either created by incident PE close to the surface or by BE leaving the sample (see figure 3.1). Both an surface that is inclined with respect to the incident primary beam and a geometrical edge of the sample result in a larger intersection of the surface with the interaction volume of the PE as compared to a flat and perpendicular surface. An increased intersection results in more BE leaving the sample and a higher generation of SE in the process. This leads to an increased electron signal on surfaces that are tilted with respect to the primary beam and on edge like structures. These effects, along with topographical features preventing electrons emitted from the surface to reach the detector, are the origin of the topography dependence of the contrast in SEM, making it possible to create images showing the geometrical configuration of samples [83, pp.144-169].

The interactions of the PE within the sample are very complex; Besides the sample surface geometry there are many other parameters that influence the image signal [85, pp.149ff.]. This includes a dependence on the atomic number of the sample material, making it possible to obtain material contrast [86] as presented in figure 3.2. Since the PE penetrate deep into the sample and both SE and BE can leave the sample again from a certain escape depth, structures below the surface may also be resolved, especially for samples with low mass density. In addition to the creation of SE, the interactions of the PE with the sample's atoms can also result in Auger electrons and electromagnetic radiation leaving the sample surface, which can be used to obtain additional information about the specimen [83, p.141].

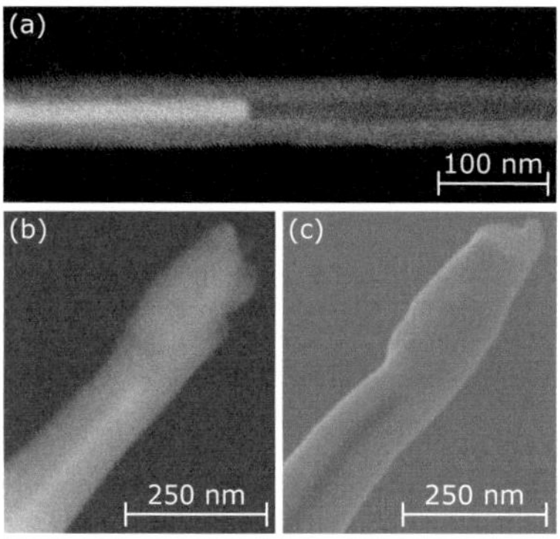

Figure 3.2: High magnification scanning electron microscopy images of iron-filled carbon nanotubes: (a) The material contrast between the iron-filled (left) and unfilled (right) part is clearly visible. (b) and (c) are images of the same iron-filled carbon nanotube that were acquired with similar settings for the detector and the primary beam but with a different geometrical arrangement within the SEM equipment. The inverted contrast of the iron filling with respect to the carbon shell between images illustrates that the relative material contrast depends on many different parameters and not the sample materials alone.

There are two main aspects that limit the spatial resolution of SEM images besides the signal-to-noise ratio: the diameter of the primary beam at the sample surface and the diffusion length of the PE inside the sample. Two objects can only be reliably resolved if they are not irradiated by the primary beam at the same time. A well calibrated field emission based electron source and electron optics can have a primary beam diameter in the order of 2 nm [87], depending on the acceleration voltage of the PEs [87].

The second limitation originates from the movement of the PE through a sample. If the lateral extent of a geometrical feature is much smaller than the possible interaction volume, any part of this feature will contribute to the signal, regardless of the position at which the primary electrons enter the sample. The feature will still be visible in this case but its actual geometrical structure can no longer be determined. The minimal feature width that can be resolved depends on the properties of the detector, the acceleration voltage of the primary beam and the material of the sample and was found to be in the order of 2 nm for high density samples and 5 nm for low density samples [87]. Both limiting factors, diameter of the primary beam at the sample surface and the diffusion length of the PE, are connected via the beam's acceleration voltage, since a higher kinetic energy will result in a more focused and therefore lower diameter primary beam but the diffusion length in the sample will increase. A minor effect further limiting the spatial resolution is given by the delocalized production of SE in a material. The contribution of this effect is usually less than 0.5 nm, however, and can be neglected [87].

To actually determine the value of the spatial resolution is difficult as the actual contrast formation is complex with many competing effects depending strongly on the properties of the sample. A theoretical limit for the smallest resolvable structure is given by the effects discussed above and the minimal feature width for low density samples of 5 nm. Additionally, the instrumentation introduces relative error on the length measurement [88]. Therefore, these two sources of error were combined to obtain the margins of error on the SEM-based length measurements in the experiments.

The SEM component of a Carl Zeiss AG '1540XB CrossBeam' workstation was utilized for every geometrical measurement in this work. The manufacturer calibrated length measurement of the instrument is supposed to exhibit a relative error of $< 1.5\%$ [88].

3.2 Focused ion beam milling

A focused ion beam instrument is pretty similar to a SEM instument in many aspects, except that its primary beam consists of accelerated ions instead of electrons. But whereas in SEM imaging a minor radiation damage to the specimen is an undesirable side effect, the focused ion beam (FIB) method is designed to be destructive. Ions are much heavier than electrons, so they can transfer considerably more momentum to the nuclei of the sample. The transferred momentum and the kinetic energy can result in atoms leaving the surface of the sample. This results in removal of the material, which can be used as a tool to modify a sample's geometry. This process is called ion beam milling and is demonstrated in figure 3.3. Not all collisions of an incident ion with the sample lead to a sample atom leaving the surface. As a side effect, the interactions between the incidental ions and the sample's material also lead to a local amorphization of the sample structure. Furthermore, the ions gradually decelerate through interactions within the sample until their movement stops, leading to a contamination of the sample with the beam material. The energy stored in the sample by the beam also leads to secondary electrons leaving the sample surface which can be collected by a detector to create an image similar to SEM. It has to be kept in mind, though, that the scanning required to create the image will damage the sample even at low beam currents [89].

The FIB instrument used throughout the experiments for this work was the already mentioned Carl Zeiss AG '1540XB CrossBeam' workstation that combines both SEM and FIB functionality with two separated beam sources and lens systems. Common to many commercially available instruments, the system uses gallium as an ion source.

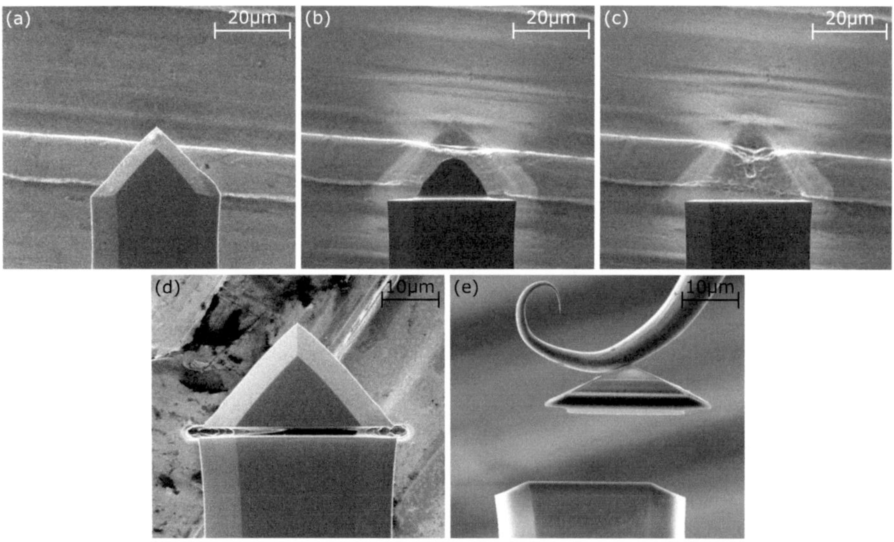

Figure 3.3: Illustration of the ion beam milling capabilities of a focused ion beam instrument: exemplary removal of the triangular free end of a picket shaped cantilever; (a)-(c) consecutive images taken during a large-area triangular shaped removal; (d)-(e) line shaped removal ('cut') followed by a micromanipulation operation

3.3 Induced material deposition and removal

A further tool to modify a micro- or nanometer scaled sample is given by electron- or ion beam assisted deposition of material. This method utilizes gaseous organic precursors containing the desired material's atoms which are injected into the instrument close to the sample. Subsequently the precursor molecules attach to the sample surface where they move by thermally activated transport mechanisms. These molecules can interact with the energetic primary beam particles as well as the back-scattered and secondary particles leaving the sample surface. Through these interactions their molecular bonds are broken or rearranged thus forming new compounds both volatile and non-volatile. The volatile species leave the sample surface and are removed by the vacuum pumps and the non-volatile species form a new material layer on the irradiated sample surface. Using an ion beam results in a more efficient material deposition process as compared to using an electron beam. In the case of using an ion beam, however, the new material is also contaminated by the ions and the destructive properties of the ion beam lead to material removal competing with the deposition. The latter requires a careful adjustment of ion beam parameters. A large variety of possible materials that can be deposited by this method is available, including metals, insulators and carbon [89, pp.75-85].

A related technique is the induced material removal, where the beam assisted dissociation of a gaseous precursor leads to highly reactive species that form volatile compounds with the sample material, thus achieving chemical etching. An example of this process is the electron beam induced etching of carbon in the presence of water molecules [90, 91].

In the vacuum conditions present in experimental setups, there is a certain residual partial pressure of hydrocarbons that can adsorb to the surfaces of solids [83, p.331]. If no preventive measures are taken [87], the material deposition process as described above, can occur unintentionally in the microscopy equipment, although to a lesser extent. During the imaging process in SEM this can lead to a contamination of samples with unwanted materials such as surface layers of carbon.

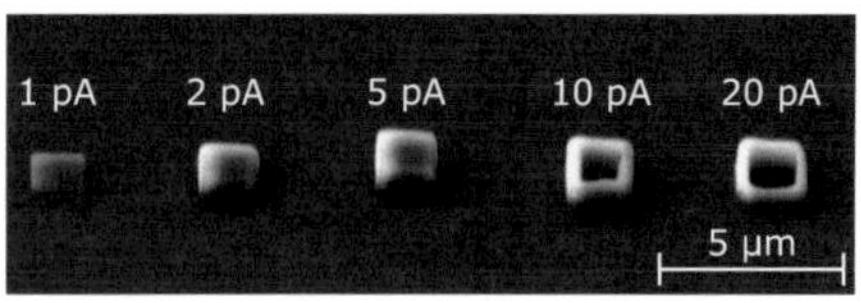

Figure 3.4: Exemplary focused ion beam induced deposition of carbon. Different ion beam currents from 1 to 20 pA were used to find the optimum balance between material deposition and removal. In this case the 5 pA beam current resulted in the most efficient material deposition.

To enable material deposition the '1540XB CrossBeam' workstation is equipped with a gas injection system that was used to provide platinum or carbon precursor gases in close vicinity to the sample. For carbon removal operations, the 'low vacuum' mode of a FEI Company 'Nova NanoSEM 200' instrument was employed. This operation mode creates a water atmosphere with a pressure in the order of one millibar inside the specimen chamber of the instrument. While the 'low vacuum' mode is intended to study biological or low conductivity samples, it can be exploited for electron beam induced etching of carbon.

3.4 Micromanipulation

In order to obtain the ability of an in-situ positioning and manipulation of the samples inside the SEM and FIB instruments' vacuum chambers, a piezo motor driven micromanipulator with a rotation tip as shown in figure 3.5 was employed[1]. The piezo motors of this system are operated by an external control unit and allow high precision movements of the tip on an approximately spherical coordinate system. The special tip further adds the possibility of an axial rotation of the cylindrical sample holders that can be plugged into it. These sample holders can be equipped with electrochemically etched sharp tungsten needles as a general manipulation tool. Alternatively, a specimen such as a dSFM cantilever can directly be attached to it. Either way, the object is glued to the sample holder using conductive silver paint[2].

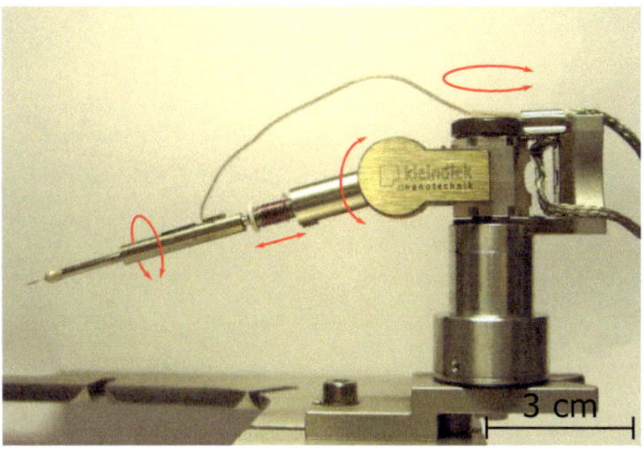

Figure 3.5: Micromanipulator with rotation tip. The movement axes are indicated. The rotation tip is equipped with a cylindrical sample holder with a tungsten needle.

In this work a typical micromanipulation process employed the above mentioned tungsten needles and had several steps: First the needle was moved into contact with another

[1]Kleindiek Nanotechnik GmbH, MM3A-EM with ROTIP-EM
[2]Acheson Silver DAG 1415

Figure 3.6: Scanning electron microscopy images of the steps of a micromanipulation operation to transfer an iron-filled carbon nanotube (FeCNT) to a new substrate: (1) approach of a tungsten needle to a substrate containing FeCNTs, (2) attachment of a FeCNT to the needle, (3) extraction of a FeCNT from its substrate, (4) approach of the FeCNT to a new sample, (5) attachment of the FeCNT to the new sample, (6) the old attachment between needle and FeCNT was broken to complete the transfer.

specimen. Then this specimen, usually an iron-filled carbon nanotube, was attached to the needle with the help of electron beam induced deposition of carbon. The next step consisted of the extraction of the specimen from its substrate by moving the manipulator. Sometimes the connection between substrate and specimen was more stable than the depositioned carbon attachment, however, which resulted in a failure of the extraction. In that case these steps were repeated with a longer electron beam irradiation time and a changed scan shape during carbon deposition to create a stronger bond until the desired outcome was obtained. After the successful extraction it was possible to transfer the specimen to another sample, if this was required. By using the micromanipulator the specimen was moved to the new sample and a contact between the desired sample feature and the specimen was established. Subsequently a connection between these two objects was again created by electron beam induced deposition of carbon, this time with a longer irradiation duration to create a more solid bond than the one between specimen and tungsten needle. Next, the manipulator was moved in order to break the first attachment thus completing the transfer. This last step was again susceptible to failure and a repetition of the last steps was sometimes necessary. Finally, the tungsten needle could be used to further manipulate the specimen if more connections between specimen and new sample were required. Images of such an micromanipulation process involving an iron-filled carbon nanotube can be found in figure 3.6.

3.5 Vibration measurements

If information about the actual mode shape of a dSFM sensor or on the resonance frequencies of substructures of the sensor are required, the data on the oscillatory behavior of a dSFM sensor acquired by deflection measurements with the dSFM equipment is insufficient. In these cases it is favorable to obtain additional data about the desired properties directly from vibration measurements inside the SEM equipment. To enable such studies, a home-build vibration stage was created with a piezoelectric shear plate actuator as the main component. Similar setups have already been used in the work of S. Philippi and U. Weißker [92, 93] but a new sample holder was designed to allow for precise positioning of dSFM sensors with low risk of damage. By applying a harmonic alternating voltage to the shear plate actuator, a specimen on the sample holder can be excited to mechanical vibration, which can be studied with the help of SEM. The vibration stage was assembled on a standard SEM sample stub and the voltage was supplied by a shielded vacuum compatible coaxial cable via a feed-through. The sensors had to be attached to the sample holder by conductive silver paint. The vibration stage and an exemplary SEM image of a vibrating dSFM sensor are shown in figure 3.7.

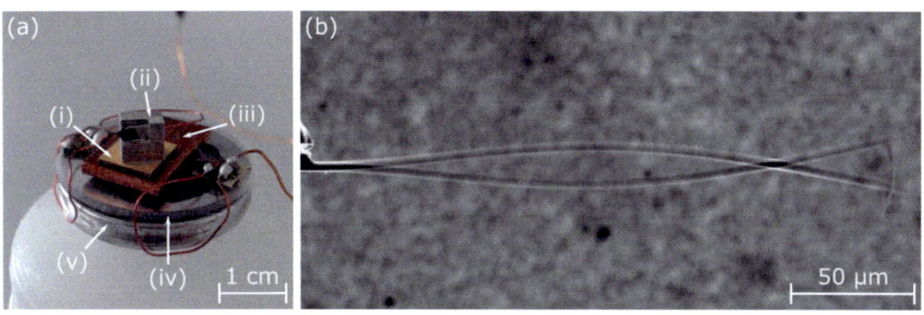

Figure 3.7: (a) Image of the vibration stage and (b) exemplary scanning electron microscopy image of a vibration measurement of the second flexural mode of a dSFM sensor. The main components of the vibration stage used for the experiments were (i) the piezoelectric shear plate actuator, (ii) the sample holder (in this case with attached cantilever carrier chip), (iii) the back electrode, (iv) insulating substrate and (v) SEM sample stub.

The oscillation frequency of typical dSFM sensors is far larger than the scan frequency of the SEM. Therefore, the images acquired by this method do not show the momentary sensor deflection. Instead, the image contrast results from an integration over many oscillation cycles of the sensor. The contrast information of a pixel is then related to the relative amount of time the sensor is at the pixel's position during these oscillation cycles. As the sensor's time of presence at the turning points of its motion is increased due to the vanishing velocity, the oscillation envelope is clearly visible.

4 Iron-filled carbon nanotubes

In most of the experiments presented in this thesis iron-filled carbon nanotubes play an essential role. This interesting composite material combines the exceptional mechanical properties of carbon nanotubes with the magnetic properties of a very high aspect ratio iron wire with a nanoscaled diameter. The FeCNTs used in the experiments presented herein have been produced by Uhland Weißker in the course of his PhD-work and their mechanical and magnetic properties have been elaborately studied in the past [92, 93, 94]. Therefore, these FeCNTs constitute a well known magnetic and mechanical sample system.

In this chapter, first a short general introduction to carbon nanotubes (CNTs) is given, followed by an overview over the production and the known properties of the FeCNTs used in this work. Then, some of the applications of this material system are introduced. This includes a summary on previous work concerning their use as magnetic interaction tips in magnetic force microscopy and their application as force sensors in SEM experiments, but also new results on their application as a magnetic test system for nanoscaled superconducting quantum interference devices (nano-SQUIDs).

4.1 Introduction to carbon nanotubes

Carbon is an element with many different solid state allotropes. These include the well known three-dimensional modifications graphite and diamond, but carbon also forms more exotic structures such as fullerenes, graphene or carbon nanotubes, which can be considered to be of lower dimensionality.

The origin of this structural diversity of carbon lies in its electronic configuration $1s^2 2s^2 2p^2$ and the low energetic difference of the 2s- and 2p-orbitals. This low energetic difference allows for an intermixing of the orbitals' wavefunctions to form hybridized orbitals. Depending on the degree of hybridization, this results in different possible coordination geometries of carbon atoms and associated structural features. Therefore, carbon can form planar structures such as the layers of graphite, and three dimensional tetrahedral structures found in diamond as well as structures of various in-between states such as fullerenes or carbon nanotubes [95, pp.6-32].

Carbon nanotubes are tubular structures of carbon which exhibit axial symmetry and yield many interesting physical properties. These include exceptional mechanical properties such as high tensile strength and a large Young's modulus at a low mass density, as well as structure dependent electronic transport behavior and many more [95, pp.186-217].

4.1.1 Structure of carbon nanotubes

Graphite consists of planar carbon layers built from regular hexagons of carbon. These constituent layers are called graphene and are bound to each other by van-der-Waals forces. Only three of the carbon's four valence electrons participate in the covalent bonding by the sp^2-hybrid orbitals necessary to form the hexagonal carbon structure. The remaining electron from each atom interacts with those of the layer's other carbon atoms to form a so-called π-electron-bonding cloud that is delocalized over the entire graphene layer. A single-walled carbon nanotube (SWCNT) can be considered an individual graphene sheet which is rolled up to form a hollow cylinder. A model of such a SWCNT is shown in figure 4.1.

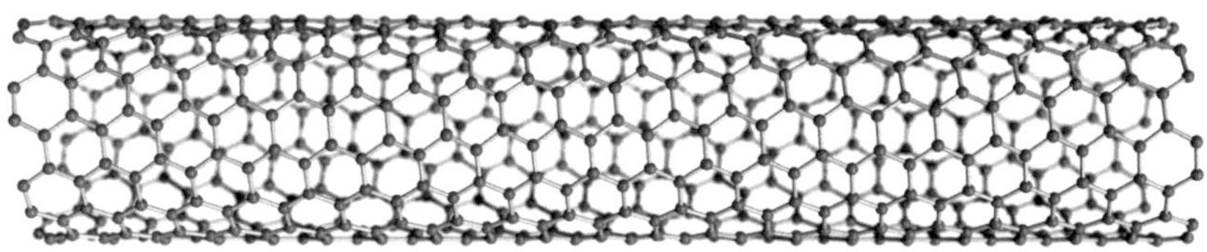

Figure 4.1: Ball-and-stick model of a single-walled carbon nanotube. The carbon atoms are colored dark gray and the bonds between the atoms light gray. The hexagon structure of the graphene sheet is visible. (Structure created with the software 'Avogadro' [96].)

The structure of a SWCNT may be classified depending on the size and orientation of the constituting rolled up graphene sheet. This results in a differentiation of zig-zig, armchair or chiral carbon nanotube structures with different diameters. The exact structure and geometry of the SWCNT strongly influences its physical properties and needs to be known when such a tube is discussed [95, pp.126-134]. The structure of an individual SWCNT can be described by its chiral vector, being a linear combination of two unit vectors of the graphene lattice [97]. Furthermore, the SWCNT's ends can be closed by additional carbon caps of various structures.

Multi-walled carbon nanotubes (MWCNTs) consist of several SWCNTs of different diameters in a concentric arrangement. Usually their interlayer distance is constant and they consist of SWCNTs with different structures. The SWCNTs in a MWCNT are not covalently bonded to each other, but interact instead only weakly by van-der-Waals forces and in some cases by their π-electron clouds [95, pp.135-139].

4.1.2 Properties of carbon nanotubes

The same way as carbon nanotubes can be considered to be rolled up graphene sheets, many of their physical properties can be derived from those of graphene with some modifications due to their geometrical structure. Likewise, the physical properties of MWCNTs result from those of the constituting SWCNTs of diverse structure and diameter with an additional contribution from the bonding between the SWCNTs.

Shortly after the first report on the discovery of CNTs [98], predictions of their exceptional mechanical properties were published [99, 100]. For SWCNTs these properties are based on the stability of the carbon bonds and the absence of a conventional crystal lattice of the rolled-up graphene sheet that constitutes a CNT [95, pp.190-193]. Importantly, no significant change of the Young's modulus by rolling up the graphene sheet is expected, except in the case of large curvatures, i.e. for small CNT diameters [101]. For MWCNTs there is an additional contribution to the mechanical properties originating from Van-der-Waals interactions between the individual tubes [102].

A first experimental demonstration of the exceptional mechanical properties was provided by measuring the thermal vibrations of several MWCNTs in a transmission electron microscope. A very high Young's modulus with an average value of 1.8 TPa [103] was determined. Later studies conducted by measuring the deflection of CNTs induced by a SFM tip in various geometries [59, 101] found lower values for the Young's modulus of multi- and single-walled CNTs, respectively. Nevertheless, these values of around one TPa correlate to the in-plane Youngs modulus of graphene within the measurement error of the experiments [95, p.191][99, 104]. Further experiments which used two SFM sensors to apply and measure a tensile load to CNTs discovered extremely high tensile strengths of up to 63 GPa [104]. The mechanical properties of the CNTs, however, are strongly influenced by defects of the carbon structure and their values might be considerably decreased for tubes with a higher amount of defects [101].

Not only the mechanical properties of CNTs have been predicted to exhibit exceptional characteristics but also their electrical properties. These include one-dimensional conduction, quantum effects and either semiconducting or metallic behavior depending on their geometry and structure [105, 106, 107, 108, 109]. Graphene is a zero gap semiconductor and the different electronic properties of SWCNTs can be regarded to arise from the periodic boundary conditions introduced by rolling up the graphene sheet [109]. By combining multiple individual SWCNTs with different structure and diameter into a concentric arrangement the resulting MWCNT usually contains metallic SWCNTs, resulting in an overall metallic behavior of the MWCNT.

For a more detailed and complete account on mechanical, electronic as well as other interesting physical properties of carbon nanotubes, a large number of journal papers, reviews and textbooks are available [95, 97, 101, 107, 109, 110, 111].

4.2 Synthesis of iron-filled carbon nanotube samples

There are several well-established production methods for carbon nanotubes available. These include arc-discarge, laser-ablation and chemical vapor deposition techniques with a large range of accessible production parameters. The methods differ in structure, purity and number of defects of the resulting carbon nanotubes as well as in their applicability to large scale production [95, pp.140-158].

The iron-filled carbon nanotubes used for the experiments within this thesis have been synthesized by chemical vapor deposition in a single step process with a two furnace setup as shown in figure 4.2. In this kind of synthesis the organometallic compound ferrocene is employed both as a carbon source and to supply part of the iron for the process. In the first furnace, a ferrocene powder is sublimated at a temperature in the order of 400 K, with the exact temperature depending on the desired mass flow of ferrocene [93]. The resulting gaseous ferrocene is then transported by a flow of argon carrier gas to the second furnace with a higher temperature. There, silicon oxide terminated silicon substrates are positioned that are coated with an aluminum interlayer and an iron catalyst layer. Upon heating, the iron layer reorders into small quasi-liquid catalyst islands [112]. These islands interact with fragments of the ferrocene precursor, which decomposes in the heat of the furnace into iron, hydrocarbons and hydrogen [113]. Given the correct process conditions the interplay between the islands and the precursor constituents then forms straight multi-walled carbon nanotubes that are filled with iron bodies of various geometries ranging from small particles to long continuous iron wires [114]. However, the synthesis process can also lead to a superficial contamination of the filled CNTs with iron or iron carbide particles [94] as well as a layer of amorphous carbon [93]. A thorough description of the process used to create the FeCNTs employed in this work may be found in the dissertation of Uhland Weißker [93].

After their synthesis, the FeCNTs are attached to the substrate material and exhibit a forest-like structure as shown in figure 4.3. The FeCNTs of the sample used for the experiments in this thesis have lengths of up to 25 μm, typical diameters between 50 and

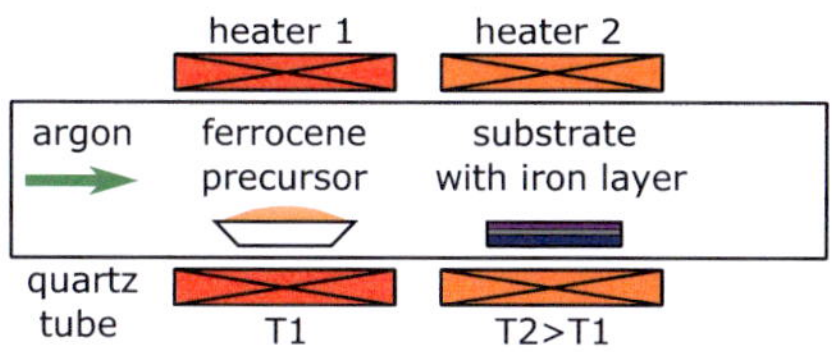

Figure 4.2: Setup used for the single step process to produce iron-filled carbon nanotubes. A furnace with two heaters creates two zones of different temperature. (according to [93])

100 nm and contain long continuous iron fillings in the diameter range of 15 to 50 nm. These iron fillings are usually wire-shaped particles, mainly consisting of single-crystalline body centered cubic α-iron [94, 112].

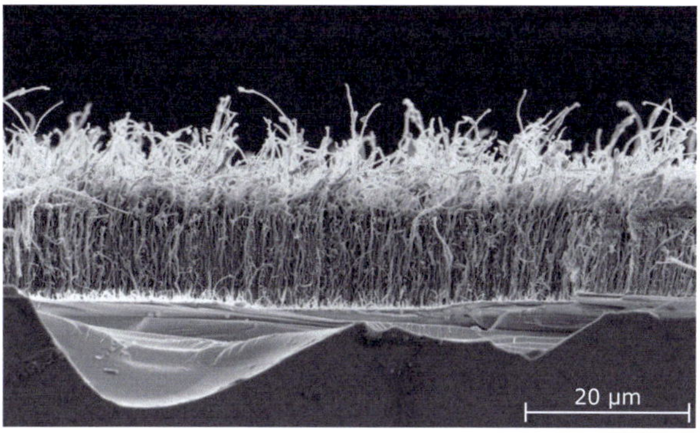

Figure 4.3: The forest-like structure of a iron-filled carbon nanotube sample. For experiments, individual FeCNTs can be extracted from the sample.

4.3 Mechanical and magnetic properties of iron-filled carbon nanotubes

The mechanical properties of iron-filled carbon nanotubes similar to the ones used in this thesis have been analyzed in various works using dynamic and static bending methods. Besides revealing values between 120 and 698 GPa for the Young's modulus of the examined FeCNTs [93, 115], the evaluation of these experiments also yielded other interesting results. First it was found that the bending mechanics of the FeCNTs can be well described by the Euler-Bernoulli beam theory discussed in section 2.3. Next, these studies indicated that the results of beam theory based evaluations can further be improved by considering the FeCNTs' attachments to their environment, which were created by electron induced deposition of carbon, as a torsional spring boundary condition [115, 6]. Beyond these results, it was found that the contribution of the iron filling to the flexural rigidity of the FeCNTs may be neglected [115]. In contrast, the mass of the iron filling provides a significant contribution to the total mass of the FeCNT. In order to calculate this mass a literature value for the mass density of iron $\rho^{Fe} = 7874\ \mathrm{kg/m^3}$ [116] at room temperature can be used.

The magnetic properties of the FeCNTs have also been examined by various experimental techniques, including magnetic force microscopy, micro-Hall magnetometry and cantilever magnetometry. First of all, these experiments indicated that the saturation magnetization of the Fe nanowire is in good agreement with literature values for α-iron [117, 118]. The saturation magnetization of α-iron is given by $M_S^{Fe} = 1.71 \cdot 10^6\ \mathrm{A/m}$ at room temperature and slightly increases for lower temperatures to $M_S^{Fe,lt} = 1.74 \cdot 10^6\ \mathrm{A/m}$ [119, 120]. Experiments also showed that the iron wires encased by the carbon usually consist of a single magnetic domain. However, a possible multi-domain state induced by strong deformations of the nanotube has been observed in MFM experiments [94]. Furthermore, it was deduced that in the iron wires inside the FeCNTs the contribution of the shape anisotropy to the magnetic energy is more than an order of magnitude higher than the maximum contribution of the magnetocrystalline anisotropy energy [118]. This results in the easy axis of the magnetization of the iron filling being aligned along the long FeCNT axis and not along any crystallographic direction. The external field necessary to reverse the magnetization of the wire-like particles inside the FeCNTs, the so called 'switching field', was observed to differ between wires of various geometries from 75mT up to 400mT [121, 94]. The magnitude of the switching fields as well as their dependence on the angle between the nanotube axis and the external field was found to be very well described by the nucleation fields of a curling mode magnetization reversal [122] and not by coherent rotation of the magnetization as would be expected from a simple Stoner-Wolfahrt model [123, p.134]. Furthermore, it was proposed that the magnetization of the iron filling is reversed by a thermally activated nucleation of a domain wall by curling mode reversal and a subsequent propagation of this domain wall through the particle [117, 118].

4.4 Applications of individual iron-filled carbon nanotubes

After the synthesis and characterization of the FeCNTs, it has been demonstrated that this composite material system, which combines the mechanical properties of a carbon nanotube with the magnetic properties of single-domain iron nanowire, is suitable for practical applications concerning magnetic field measurements.

4.4.1 Tips for magnetic force microscopy sensors

One application of the FeCNT is their use as tips for MFM sensors [5] by attaching them to the already existing tip of a SFM sensor as shown in figure 4.4. These magnetic FeCNT tips have several advantages when compared to conventional MFM tips.

Carbon nanotubes have been employed as tips for scanning force microscopy sensors before due to their advantageous properties: the high aspect ratio of carbon nanotubes reduces image artifacts when measuring samples with steep edges and their elastic properties help to protect both the nanotube tip and the sample from damage during the measurement process [124, 125]. Furthermore, if single-walled carbon nanotubes are employed, their small diameter also results in extremely sharp tips with lateral dimensions in the order of single molecules enabling scanning force microscopy imaging with very high lateral resolution [126]. However, the elastic properties of carbon nanotubes can also cause image artifacts, in particular their low stiffness with respect to bending deformations [127, 128]. In addition to this, carbon nanotubes can be employed as templates to define the shape of magnetic [129] or conductive material [130] in order to create high aspect ratio specialized SFM tips for various applications. For this, the material is usually coated on the surface of the CNT, making these special tips less durable.

A practical advantage of the FeCNT-based MFM tips is given by the mechanical as well as chemical stabilization of the iron nanowire by the carbon shell. Commercially available MFM sensors are normally standard microfabricated dSFM sensors made from silicon with the sharp silicon tip coated by a layer of ferromagnetic material [71]. As this coating constitutes the outer surface of the tip it can be damaged during the MFM measurement process by contact with the sample material, thus changing its geometrical and magnetic structure. The same applies to the above mentioned CNT tips coated with magnetic material. In the case of FeCNTs, however, the magnetic material is protected by the durable carbon shell, thus ensuring constant geometric and magnetic properties. In addition to this, the carbon prevents oxidation of the iron core even when stored under ambient conditions.

The other advantages of the FeCNT tips are based on the favorable magnetic properties of the iron nanowire: it has been proposed before that high aspect ratio single-domain ferromagnetic nanowires with high saturation magnetization would constitute ideal MFM tips [66, 131]. Such long wires can be considered an elongated dipole with the two poles

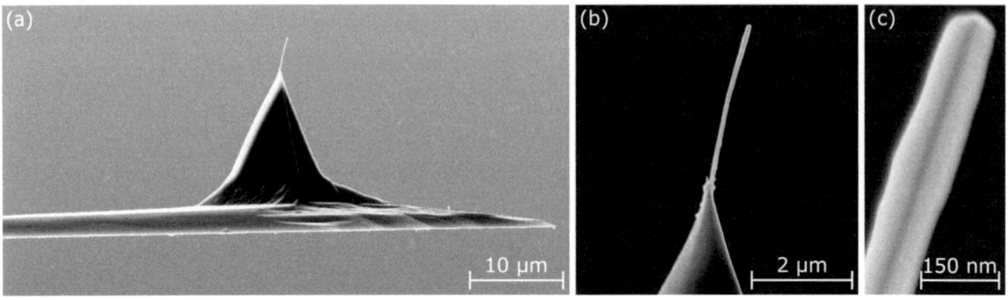

Figure 4.4: Iron-filled carbon nanotube based magnetic force microscopy sensor: (a) view of the end of the cantilever with FeCNT attached to the silicon tip, (b) close up of the silicon tip's apex and the FeCNT, (c) high magnification image of the nanotube's free end with the material contrast of the iron filling clearly visible.

separated by the length of the wire. If this separation is far longer than the decay length of the magnetic field of the sample only the pole closest to the sample would contribute to the magnetic MFM signal, thus behaving monopole-like. The same reasoning can be applied to the point probe approximation introduced in section 2.5.2. The parameters of this approximation, the effective monopole moment q and its position within the tip structure, are determined by the magnetic surface charges of the tip within the sample's magnetic stray field [68]. For a long single-domain ferromagnetic nanowire with a sufficiently flat end, all of of the surface charges close to the sample would be subjected to approximately the same magnetic stray field as long as the stray field decay length is much larger than the diameter of the nanowire. In contrast, the surface charges at the other end would not be immersed into the field at all. Hence, the amount of the surface charges within the stray field, and connected to this the parameters of the monopole point probe approximation, are independent of the stray field geometry. This is a significant improvement over conventional MFM tips as for these tips the effective magnetic volume and therefore their monopole point probe approximation parameters depend on the geometry of the sample stray field [71].

By using micron-sized conducting lines to create magnetic stray fields of various geometries, it has been demonstrated that the magnetic properties of FeCNT MFM tips are well described by a pure monopole point probe approximation. Furthermore, the parameters of this approximation indeed exhibited an independence of the magnetic stray field of the sample as expected for long single-domain ferromagnetic nanowires [5]. Therefore, the frequency shift signal of MFM sensors based on such FeCNT tips may be described by equation 2.45 for a pure monopole point probe for any magnetic sample stray field. The effective monopole moment q and its position within the FeCNT may be obtained either from calibration MFM measurements or an approximation can be derived from geometrical data of the FeCNT [132].

4.4.2 Cantilever magnetometry

Cantilever magnetometry is a method to study the magnetic properties of small magnetic samples by placing them on a cantilever beam and resonantly oscillating the system in an well defined homogenous external magnetic field. The magnetic properties of the sample can be derived from the response of the oscillatory state of the cantilever to changes of the external magnetic field. In this aspect this measurement principle is reminiscent to MFM but due to the homogeneity of the external field the frequency shift in cantilever magnetometry is not induced by a field gradient but rather by the restoring torque from the magnetic interaction of the sample with the external field [133]. To compensate for the comparatively low force gradients induced this way by small magnetic samples, long and thin micro-fabricated silicon cantilevers with low spring constants and resonance frequencies are employed [133, 134]. In order to suppress the thermal vibrations of the cantilever, often measurements at low temperatures are required to increase sensitivity [118].

Instead of using a separate cantilever, the FeCNTs can be employed directly as both the cantilever beam and magnetic sample. By studying their oscillatory behavior in the magnetic field of small spherical permanent magnets inside the SEM it has been demonstrated that this cantilever magnetometry setup results in a much higher signal and can be utilized at room temperature [6]. The high signal is based on the FeCNT exhibiting a low spring constant while at the same time being very short. The latter is advantageous as a longer cantilever reduces the signal in cantilever magnetometry [133]. The results from this experiment can be adapted for the study of magnetic properties of arbitrary nanoscaled magnetic samples if the iron-filled carbon nanotube is replaced by an empty one and the magnetic sample is placed at the CNT's free end. The measurement of the oscillation of CNTs, however, usually requires complex detection methods like SEM, which limits the applicability of this method [6].

4.4.3 Test samples for nanoscaled magnetometer devices

The FeCNTs constitute an excellent magnetic nanoparticle test sample to verify the functionality of magnetic measurement devices due to their well studied magnetic properties, their mechanical and chemical stability and available experience concerning their micromanipulation. In the following new experiments, a FeCNT was used to demonstrate the applicability of low-noise nanoscaled superconducting quantum interference devices (nano-SQUIDs) for practical studies of magnetic nanoparticles [135][1].

Low noise nanoscaled superconducting quantum interference devices

Superconducting quantum interference devices (SQUIDs) make use of two superconducting phenomena: flux quantization in a superconducting ring and the Josephson effect. They can be used as magnetometers to measure the relative change of magnetic fields with very high precision [136, p.63]. A common implementation of such a device is the direct current SQUID in which a loop of superconducting material contains two Josephson junctions. These Josephson junctions consist of a 'weak link' in the superconducting material and can be created by different methods including tunnel barriers, normal metal layers by making use of the proximity effect, narrow geometrical constrictions, grain boundaries [137] and others [138, p.197]. An account on the basic principles of SQUID magnetometers is available from basic textbooks on superconductivity [136, 138] as well as dedicated literature [139].

Micro- or nanoscaled SQUIDs can measure the changes of magnetic stray flux of closely placed magnetic nanoparticles that couple into the SQUIDs. By applying an external magnetic field this can be used to study the particles' magnetization reversal behavior [140]. A further reduction of the size of the SQUID's loop and its Josephson junctions is required to enhance the coupling of the device to the magnetic stray field of smaller spin systems [141], to reduce the flux-noise and to improve the device performance in high magnetic fields [142]. The low noise nano-SQUIDs that were used in the following experiments were based on grain boundary Josephson junctions in FIB patterned $YBa_2Cu_3O_7$ loops [142] with an additional geometrical constriction (see figure 4.6) that allows for an optimization of the coupling of a magnetic nanoparticle's flux into the nano-SQUID [143]. A detailed description of the fabrication, characterization and modes of operation of the low noise nano-SQUIDs used for the experiments is available elsewhere [135].

Placing an iron-filled carbon nanotube onto a nanoscaled superconducting quantum interference device

As the objective of this experiment was to use an individual FeCNT as a magnetic test nanoparticle for the nano-SQUIDs, the FeCNT had to be added to the device. First an FeCNT was extracted from the FeCNT sample substrate by using a tungsten needle

[1] This section contains new results from a cooperation with the group of Dieter Kölle (Universität Tübingen). The fabrication of the nano-SQUIDs as well as all the nano-SQUID related calculations and low temperature measurements and corresponding data, including the derivation of the factor ϕ_M are courtesy of Tobias Schwarz and his colleagues from this group.

according to the standard micromanipulation procedure as discussed in section 3.4. To study magnetic nanoparticles with nano-SQUIDs, an external magnetic field has to be applied. This applied external field has to be oriented parallel to the substrate surface of the device and perpendicular to its grain boundary plane in order to reduce the coupling of this field into the loop of the nano-SQUID and to avoid the suppression of the critical current in the grain boundary Josephson junctions [142]. This was a strong constraint for the positioning of a FeCNT, since the easy axis of its iron core, corresponding to the tube's long symmetry axis had to be oriented as parallel as possible to the external magnetic field in order to obtain small switching fields and to reduce the influence of measurement errors of the angle between the external field and the easy axis on these switching fields [117]. Additionally not only the alignment of a FeCNT but also its position with respect to the nano-SQUID's constriction was important. Generally, the optimal coupling of an arbitrary magnetic nanoparticle's stray field to a nano-SQUID is achieved by positioning it on top of this constriction [143]. The FeCNTs, however, exhibit a comparatively strong magnetic stray field which may be above the upper detection limit of the sensitive nano-SQUID's measurement electronics. Hence a certain separation of a few hundreds of nanometers between the FeCNT's end and this constriction was required to decrease the coupling.

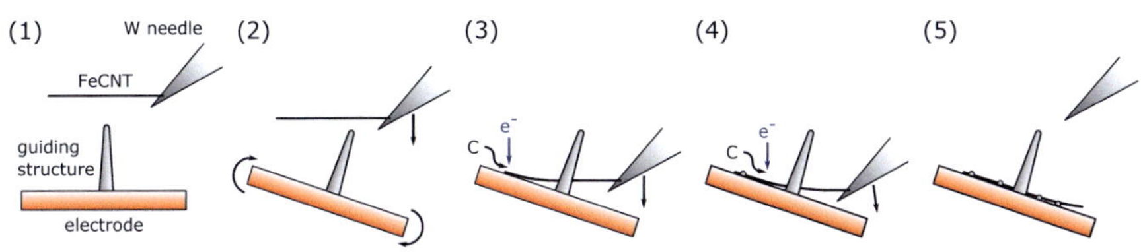

Figure 4.5: Micromanipulation steps used to align the iron-filled carbon nanotube to an electrode surface: (1) starting point: electrode with carbon guiding structure and FeCNT on a micromanipulator tungsten needle, (2) the substrate was tilted and the FeCNT was approached using the guiding structures, (3) the FeCNT's free end was pressed on the electrode surface and attached by electron beam induced carbon deposition, (4) the tungsten needle was moved to press additional parts of the FeCNT on the substrate and to secure them with carbon, (5) completed alignment with removed tungsten needle.

To account for the specific alignment requirements, the remainder of the micromanipulation operation was more complex than the standard procedure (see figure 4.5): Before the positioning of the FeCNT on the nano-SQUID, pillar-like guiding structures were created by electron induced carbon deposition on one of the SQUID's electrodes to ensure an alignment perpendicular to the grain boundary direction. Then the SQUID substrate was tilted slightly by an angle of 9° to facilitate a parallel orientation of the FeCNT with respect to the SQUID surface later on. Next, the FeCNT was approached to the SQUID while the guiding structures were used to mechanically guide the FeCNT to ensure optimal alignment along this direction. After this step, the needle with the FeCNT was moved downward to press the free end of the FeCNT on the electrode. By exploiting the elastic properties of the CNT, an alignment of the free end approximately parallel to the electrode surface was achieved. As the FeCNTs free end was now aligned along the required direction it was secured to the electrode surface by electron beam induced deposition of carbon. After this first attachment the needle with the other end of the FeCNT was moved further down and backward to put another part of the FeCNT on the electrode and fasten it with carbon. This process was repeated several times until the whole FeCNT was fixed to the electrode as flat as possible. Then the tungsten needle was broken free from the FeCNT to complete the transfer and to obtain the completed sample presented in figure 4.6.

SEM images of the completed nano-SQUID - FeCNT combination revealed that the nanotube in this experiment contained a single continuous nanowire with a length of $l_{Fe} \approx 13.8\ \mu\text{m}$ and diameter of $d_{Fe} = 39 \pm 5$ nm. The precise positioning resulted in a deviation angle of less then 1° with respect to the direction perpendicular to the grain boundary. Further SEM images from a different angle allowed for a deduction of the inclination angle with respect to the electrode surface and yielded a value of $\theta < 4^\circ$.

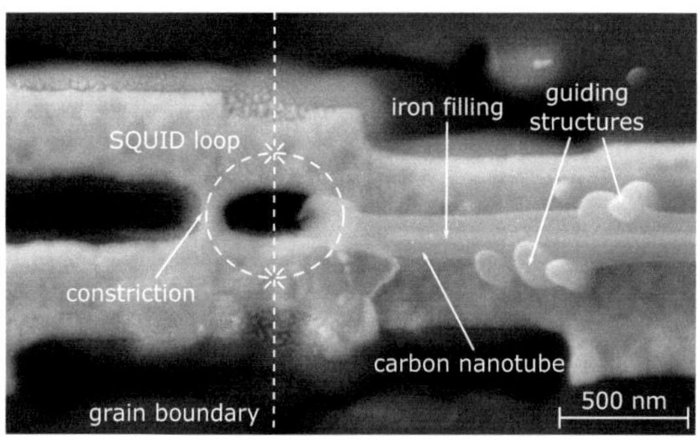

Figure 4.6: Nanoscaled quantum interference device with an iron-filled carbon nanotube. The grain boundary and the SQUID loop with its grain boundary Josephson junctions are indicated.

Calculation of the switching field of the iron nanowire

As mentioned above, the magnetic switching behavior of FeCNTs similar to the one used in this experiment can be well described by a curling mode magnetization reversal. The nucleation field H_n of a ferromagnetic wire at which the curling of the magnetization starts can be written as [117, 122]:

$$H_n = \frac{M_S}{2} \frac{\eta(1+\eta)}{\sqrt{\eta^2 + (1+2\eta)cos^2(\theta)}} \approx \frac{M_S}{2}\eta \tag{4.1}$$

with the approximation valid for small angles. The definition

$$\eta = -1.08(\frac{2\lambda_{char}}{d})^2 \tag{4.2}$$

contains the inverse ratio of the diameter d of the wire to the characteristic length $\lambda_{char} = \sqrt{4\pi A/(\mu_0 M_s^2)}$ of the material [122, 144]. This quantity contains the exchange constant A and is directly related to the exchange length of a material $\lambda_{exc} = \lambda_{char}/\sqrt{4\pi}$ below which the energy contribution of the exchange interaction of a magnetic sample is larger than its stray field energy [145]. Please note that in literature several definitions of the exchange length coexist: sometimes the characteristic length and the exchange length are even used interchangeably and the correct denomination and the actual definition are subject to discussion [67, 146].

As the inclination angles of the FeCNT on the nano-SQUID were found to be very small, equation 4.1 could be used to calculate the switching fields. With the low temperature saturation magnetization of iron $M_S^{Fe,lt}$ and the characteristic length $2\lambda_{char} = 12$ nm from previous low temperature experiments [117], a switching field of $\mu_0 H_n = -112 \pm 29$ mT was derived for the FeCNT. The error mainly originated from the uncertainty of the diameter measurement of the iron filling.

Magnetic flux measurements

The data that resulted from a measurement of the FeCNT's stray flux using the above shown nano-SQUID is presented in figure 4.7. The experiment was conducted at low temperatures of $T = 4.2$ K and with the external field aligned as described above and a field sweep rate of $\mu_0 \partial H/\partial t \approx 1$ mT/s [135]. The flux signal showed an abrupt change at fields $\mu_0 H_n \approx \pm 101$mT which corresponds well to the above calculated theoretical switching field within its error. The slight slope of the signal and the hysteresis above and below the abrupt change can be attributed to a coupling of the external field to the SQUID and to other signals from the environment, respectively [135]. The hysteresis in the signal nicely describes the magnetization reversal of a single domain particle as was expected for the FeCNT.

It is possible to recalculate the saturation magnetization of a magnetic nanowire into a SQUID saturation flux signal Φ_S [147]. The required numerical approach includes the derivation of the coupling factor of the nano-SQUID for point-like magnetic particles at a certain position and integrating over all positions in the volume of a ferromagnetic nanowire [147] as derived from geometrical data from SEM images. This results in a factor ϕ_M that relates the flux signal to a magnetization of the wire along its long axis.

From such a calculation a factor of $\phi_M \approx 47.6\ \text{n}\Phi_0/(\text{Am}^{-1})$ has been obtained for the iron filling of the FeCNT on the nano-SQUID with Φ_0 being the magnetic flux quantum [135]. Combining this result with the low temperature saturation magnetization of iron $M_S^{Fe,lt} = 1.74 \cdot 10^6$ A/m the resulting saturation flux signal was found to be $\Phi_S = \phi_M M_S^{Fe,lt} \approx 82.8\ \text{m}\Phi_0$. This also corresponds well to the observed flux signals of $\Phi(H = 0) \approx \pm 82.5\ \text{m}\Phi_0$ at zero external field [135][2]. In summary, this experiment demonstrates

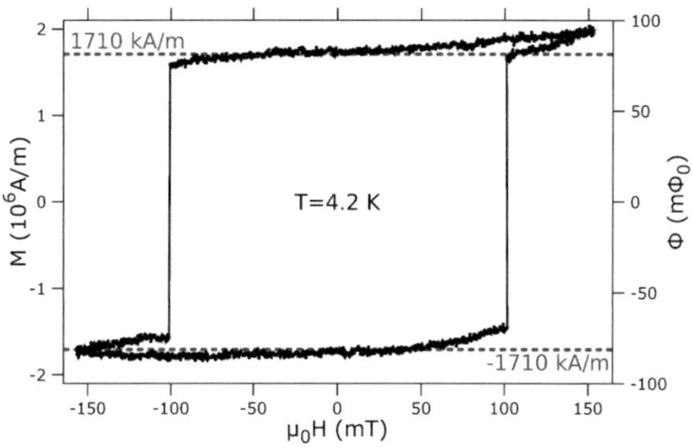

Figure 4.7: Flux signal versus external magnetic field measured with the nanoscaled quantum interference device. The axis on the left-hand side was recalibrated to show the corresponding calculated magnetization of the iron wire. The dashed lines indicate the value of the saturation magnetization of iron used in the corresponding publication [135] which is slightly lower than the value used for the calculations in this work. (Inspired by [135])

that the nano-SQUIDs and corresponding evaluation methods can measure the switching field and saturation magnetization of the FeCNT with high precision. This substantiates the applicability of the nano-SQUIDs to the study of magnetic nanoparticles and provides a further reaffirmation of the magnetic properties of the FeCNTs.

[2]The calculated values for the switching field and the saturation flux signal differ by a small amount from the ones in the publication [135] as slightly different low temperature literature values for the saturation magnetization and characteristic length were used in this thesis for consistency. Of course the general results remained the same.

5 Bidirectional scanning force microscopy

In standard dSFM the interaction tip oscillates in a direction approximately perpendicular to the sample surface causing it to be sensitive to force gradients in this direction. In some cases, however, a lateral force gradient sensitivity is desired, which requires an oscillation of the sensor's tip in a lateral direction. Recently a novel bidirectional sensor concept was developed [4, 148], which provides a sensitivity that is switchable between the perpendicular and a lateral direction by using two different resonance frequencies and the corresponding flexural modes of the sensor's cantilever beam. The topic of this chapter is to discuss this sensor concept, to describe the fabrication of corresponding sensors and to verify their viability for quantitative force gradient studies. This chapter starts with a short review of multi-modal dSFM methods in order to classify the sensor concept.

5.1 Introduction to multi-modal and lateral scanning force microscopy

The new bidirectional sensor approach is based on the excitation at the first and second flexural resonance frequency of a microstructured cantilever to measure the standard perpendicular as well as a lateral gradient of a force field. Nowadays, many different dSFM techniques have been developed that involve either the excitation of multiple frequencies or employ another resonance mode of the dSFM sensor besides the fundamental one typically excited in standard dSFM. Many dynamic mode lateral force microscopy techniques that exhibit a sensitivity to lateral force gradients are found among the latter.

5.1.1 Multi-modal scanning force microscopy

Even if the cantilever is excited at its fundamental resonance frequency, non-linear components of the tip-sample interaction force result in a contribution of higher harmonics to the sensor's motion. By measuring the amplitudes of these higher harmonics it is possible to derive the actual tip-sample force within the oscillation range of the fundamental mode [149]. These higher harmonics are related to fast decaying higher derivatives of the tip-sample force and can therefore be used to improve the spatial resolution of a dSFM measurement [150]. The amplitudes of these higher harmonics are usually very small, so it is advantageous to use special cantilever geometries that are designed to match the frequencies of specific higher harmonics with those of higher order flexural modes of the cantilever. This results in a resonant amplification of the amplitude of these harmonics, which simplifies their measurement [151].

Beyond measurements employing the first flexural mode, it has been demonstrated that the increased effective spring constant of higher flexural modes can help stabilize a very low amplitude oscillation of a dSFM sensor to facilitate obtaining atomically resolved topography data from a sample surface [152]. Furthermore, the excitation of different higher modes of dSFM sensors can be employed to oscillate the sensor's tip in a direction different from the standard vertical one. This can be exploited to implement LFM as will be discussed in more detail below.

In bimodal dSFM the fundamental and a higher flexural resonance mode of a standard dSFM sensor are both excited at the same time. This can be employed for a simultaneous measurement of the topography and a gathering of information on a different material property by detecting the oscillatory parameters of both modes. This approach has been applied to compositional contrast [153, 154] and, as already mentioned before in section 2.5.3, magnetic interactions [73, 74, 155] to avoid some of the disadvantages of the two-pass technique.

5.1.2 Dynamic mode lateral force microscopy

In order to implement dSFM measurements of lateral force gradients above a sample, the interaction tip of the dSFM sensor has to be oscillated on a trajectory approximately parallel to the sample surface. This kind of dSFM method is applied to study the lateral forces involved in atomic manipulation operations [156], friction effects close to a sample surface [157] or lateral magnetic stray field components [158].

Lateral force gradient sensitivity can be achieved in a straightforward way by rotating a dSFM probe with an appropriately placed tip by an angle of 90° as shown in figure 5.1(a). This geometrical setup is then called pendulum geometry and has been used for lateral force gradient and friction measurements with atomic resolution [157] as well as to perform measurements sensitive to lateral force gradient components of magnetostatic interactions [158, 159]. This method has the further advantage that any forces pulling the cantilever towards the sample will act in the very stable axial direction of the cantilever. Therefore, the 'jump-to-contact' instability can be avoided and very sensitive low-stiffness cantilevers may be used [160]. In this setup the dSFM sensor is fixed to this position, however, and cannot easily be switched to perpendicular sensitivity.

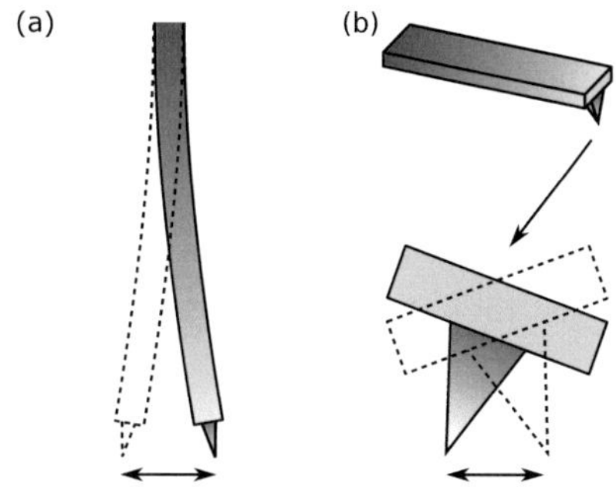

Figure 5.1: Established lateral force microscopy methods: (a) pendulum geometry and (b) torsional resonance mode

A second method to achieve a lateral motion of the tip of a dSFM sensor is by exciting a torsional vibration of the cantilever as depicted in figure 5.1 (b). This torsional resonance mode is compatible with conventional dSFM sensors and can still be switched to perpendicular sensitivity by exciting the first flexural mode of the cantilever instead of a torsional mode [161]. It has also been proven to be applicable to magnetic stray field measurements [162, 163]. This method relies on non-standard excitation and detection schemes [164], however, and therefore cannot be used with all dSFM equipments.

These two methods, the pendulum geometry and the torsional resonance mode, are well established. There are also other possibilities to implement lateral dSFM measurements. One such possibility is given by using the lateral vibration mode of the cantilever, but this vibration mode is usually not detectable by common dSFM equipment and is therefore difficult to use for dSFM measurements. Regardless, an experiment has been reported that exploited asymmetries of the cantilever to excite and measure the lateral resonance frequency of this vibration mode [165]. There have also been approaches that include the use of complex sensor geometries to achieve lateral force gradient sensitivity. A report on such a sensor utilized a cantilever-in-cantilever geometry with two microfabricated hinges and a combined mechanical and magnetic excitation mechanism to simultaneously measure perpendicular and lateral force gradients [166].

A novel method for implementing dynamic mode lateral force microscopy is given by the bidirectional sensor concept which is the topic of this chapter and will be discussed in more detail in the next section. This concept employs sensors with a special geometry and the second flexural oscillation mode of the cantilever to achieve lateral sensitivity while still retaining the possibility to conduct standard perpendicular dSFM experiments using the first flexural mode. After the publication of the first results on magnetic force microscopy experiments with this bidirectional sensor concept [148], a theoretical study proposing a similar concept was presented. This concept is based on the first and second mode flexural beam oscillations of a quartz beam clamped on two sides to give rise to perpendicular and lateral tip oscillations of the free end of an attached wire [167].

5.2 Bidirectional sensor concept

The novel bidirectional sensor concept is based on a special sensor geometry and corresponding mode of operation, as shown in figure 5.2. The interaction tip of the sensor is not placed at the end of the cantilever beam but on a stiff spacer element that is positioned on the nodal point of the second flexural mode of the cantilever instead. When the cantilever is excited at its second flexural resonance frequency the nodal point exhibits no vertical movement but instead an oscillating change of slope. The stiff spacer element translates this change of slope into a lateral motion of the interaction tip resulting in sensitivity to lateral force gradients according to equations 2.11 and 2.12 as discussed in section 2. The sensor still retains the perpendicular force gradient sensitivity of a standard dSFM sensor when its fundamental flexural mode is excited [4, 148]. Please note that both oscillation trajectories also include components of the respective other direction. For the oscillation amplitudes used in dSFM, however, these components are very small and therefore negligible.

5.3 Dynamic effective spring constants

A prerequisite for quantitative evaluation of frequency shift data obtained with any dSFM sensor is the knowledge of its effective dynamic spring constant. In the case of a bidirectional sensor that is operated according to the concept presented in the previous section, two dynamic spring constants are required, one for each operation mode. In section 2.3.3 it was shown that expressions for the dynamic effective spring constants for the first two flexural modes of a cantilever $k_{1,L}^{dyn}, k_{2,L}^{dyn}$ can be obtained. However, these spring constants are only valid for force gradients acting on the free end of the cantilever $x = L$.

Since the bidirectional sensor design requires the interaction element to be placed at the nodal point of the second flexural oscillation, this needs to be included in the calculation of the spring constants. The nodal point of the second flexural mode for a cantilever with

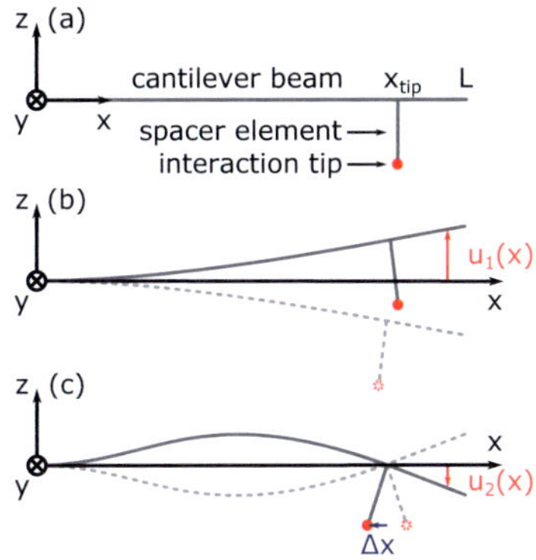

Figure 5.2: Schematic of the bidirectional sensor concept: (a) geometrical setup and components, (b) excitation of first flexural mode for common out-of-plane force gradient sensitivity, (c) excitation of second flexural mode for lateral force gradient sensitivity.

uniform cross section can be found in the framework of the Euler-Bernoulli beam theory by solving for the zero values of its modal shape $u_2(x)$ in the interval $(0, L]$ numerically. Such a calculation yielded the position $x_0 = x_{tip} \approx 0.783L$ of the nodal point as a constant fraction of the cantilever's length L.

The effective dynamic spring constants are defined in analogy to a spring in a simple spring mass oscillator. By this analogy, the energy stored in a cantilever may be written as the maximum elastic energy $U_{max}^{elastic}$ in the turning point of its motion at which the kinetic energy equals zero. Thereby a dynamic spring constant $k_{x,n}^{dyn}$ for any position x along the length of the cantilever can be defined in analogy to the harmonic oscillator by [45]:

$$U_{max}^{elastic} = \frac{1}{2} k_{x,n}^{dyn} \left(u_n(x)\right)^2 \tag{5.1}$$

with $u_n(x)$ being the shape, i.e. the vertical displacement amplitude of the neutral axis of the cantilever for the flexural vibration mode n as can be calculated by the Euler-Bernoulli beam theory. The maximum elastic energy $U_{max}^{elastic}$ of the cantilever is independent of the point of reference for the definition of the spring constant. Therefore, it is possible to relate the spring constants for different positions by using the right hand side of equation 5.1. Thus, an expression to calculate the dynamic spring constant of the first flexural mode with the position x_{tip} of the interaction tip as a point of reference

$$k_{x_{tip},1}^{dyn} = \left(\frac{u_1(L)}{u_1(x_0)}\right)^2 k_{L,1}^{dyn} \tag{5.2}$$

can be derived based on the spring constant for a tip at the end of the cantilever $k_{1,L}^{dyn}$. Since x_{tip} is a constant fraction of L for a beam with an uniform cross section, a constant factor $u_1(L)/u_1(x_{tip}) \approx 1.42$ was calculated numerically by utilizing equation 2.23. Combining this with the relationship between the dynamic spring constant $k_{1,L}^{dyn}$ and the static spring constant of the cantilever k_L^{stat} as given by equation 2.33, this results in a very simple approximation for the required dynamic spring constant $k_{x_{tip},1}^{dyn}$:

$$k_{x_{tip},1}^{dyn} \approx 2.09 k_L^{stat}. \tag{5.3}$$

A similar way was employed to derive an expression for the respective dynamic spring constant of the second flexural mode corresponding to a lateral oscillation of the interaction tip. Here, the length of the spacer element had to be taken into account.

Under the assumption of a rigid spacer element, the lateral displacement amplitude of the interaction tip Δx was derived by geometrical considerations: The deflection angle θ of the spacer element is connected to the slope of the cantilever at the position of the spacer element by $\tan\theta = u_2'(x_{tip})$. Using basic trigonometry this leads to an expression for the oscillation amplitude of the spacer element's free end:

$$\Delta x = l_{dist} \sin\left(\arctan(u_2'(x_{tip}))\right) \approx l_{dist} u_2'(x_{tip}). \tag{5.4}$$

Here, the effective spacer length l_{dist} denotes the distance between the apex of the interaction tip and the neutral axis of the cantilever. The small angle approximation on the right hand side of this equation is valid for small amplitudes, as are usually used in dSFM.

As the mode shape of the cantilever is unambiguously defined by the lateral displacement of the interaction tip Δx, equation 5.1 can be adapted to include this lateral displacement amplitude with the same arguments as given above:

$$U_{max}^{elastic} = \frac{1}{2} k_{\Delta x,2}^{dyn} (\Delta x)^2. \tag{5.5}$$

The maximal elastic energy of the system is again independent on the point of reference of the spring constant. Therefore, the dynamic spring constant for the lateral oscillation can be related to the dynamic spring constant with respect to the free end of the cantilever. By combining equations 5.1, 5.4 and 5.5 an expression for the dynamic spring constant of the second flexural mode for point-like lateral forces acting on the interaction tip can be derived:

$$k_{\Delta x,2}^{dyn} = \left(\frac{u_2(L)}{l_{dist} u_2'(x_{tip})}\right)^2 k_{L,2}^{dyn}. \tag{5.6}$$

The derivative of the mode shape is proportional to β_2, which is in turn inversely proportional to the length L of the cantilever. Therefore, the linear relationship $u_2(L)/u_2'(x_{tip}) \approx 0.240L$ was calculated from the Euler-Bernoulli beam theory. With these results and the relationship between the dynamic spring constant of the cantilever in the second flexural mode and its static spring constant as given by equation 2.35, a good approximation for the relation of the dynamic spring constant of the lateral operation mode of the sensor to its static spring constant can be found:

$$k_{\Delta x,2}^{dyn} \approx 2.31 \left(\frac{L}{l_{dist}} \right)^2 k_L^{stat}. \tag{5.7}$$

With these calculated dynamic spring constants for the two operation modes of the sensor, perpendicular and lateral, it is possible to recalculate experimental frequency shift data obtained in the perpendicular or lateral operation mode into tip-sample force gradient data by using equations 2.11 and 2.12 resulting in

$$\Delta f_z = -\frac{f_{0,1}}{2k_{x_{tip},1}^{dyn}} \frac{\partial F_z}{\partial z} \left(\approx -\frac{f_{0,1}^*}{2k_{x_{tip},1}^{dyn}} \frac{\partial F_z}{\partial z} \right) \tag{5.8}$$

$$\Delta f_x = -\frac{f_{0,2}}{2k_{\Delta x,2}^{dyn}} \frac{\partial F_x}{\partial x} \left(\approx -\frac{f_{0,2}^*}{2k_{\Delta x,2}^{dyn}} \frac{\partial F_x}{\partial x} \right) \tag{5.9}$$

with z denoting the perpendicular and x the lateral direction in accordance with the coordinate systems defined in figure 5.2. The approximations in the brackets are valid in the case of a low damping of the corresponding cantilever mode.

If the assumption of a constant force gradient over the oscillation amplitude is not justified, more elaborate deconvolution methods as mentioned in chapter 2 can be used.

5.4 Fabrication of a bidirectional sensor

To create a sensor based on the bidirectional concept presented in section 5.2 three main components are needed: a cantilever, a spacer element and an interaction tip. The exact realization of these components may vary according to the intended application of the sensor and available production facilities. In this work, commercially available silicon cantilevers, spacer elements created by focused ion beam induced depositon of carbon and FeCNT magnetic interaction tips were utilized in order to fabricate bidirectional sensors sensitive to magnetostatic force gradients. It is important to note that the following production steps that were used to prepare and assemble these components are also applicable to bidirectional sensors for different force types, assuming that a different tip is used.

5.4.1 Preparation of the cantilever

The fabrication of one of the bidirectional sensors used in this work always started with a tipless cantilever as a basic component. These cantilevers were picket shaped as can be seen in figure 5.3(1). The triangular free ends were removed from the cantilevers by using focused ion beam milling (see figure 5.3(2)). This resulted in approximately constant cross section areas along the cantilever's length allowing for the use of the Euler-Bernoulli beam theory to calculate the required position of the nodal point $x_0 \approx 0.783L$ as a given fraction of the cantilever's length. This step is not mandatory but it facilitates the determination of the nodal point of the second flexural mode of the cantilever and ensures that the derived spring constants are applicable for the fabricated sensor. For more complex cantilever shapes, it is also possible to calculate the position of the nodal point [42, 168] or determine it by vibration experiments as already shown in section 3.5. In such a case, however, the expressions for the spring constants have to be also adapted. Please note that for a cantilever as three-dimensional structure, the nodal point of the two-dimensional mode shape derived by the beam theory translates into a nodal line across its width.

5.4.2 Fabrication of the spacer element

The next step in the preparation process was to add a spacer element at the calculated position of the nodal line. After measuring the cantilever's length L, the position of the nodal line was determined at a distance of $(L - x_{tip}) = 0.217L$ from the free end of the cantilever and marked by using the focused ion beam to cut a small position mark (see figure 5.3(3)).

The spacer element is supposed to be acting as a lever to translate the oscillating change of the cantilever's slope at the nodal line into an in-plane motion of the interaction tip placed at its free end. Therefore, this spacer element is required to be rigid, e.g. its resonance frequency should far exceed that of the second flexural mode of the cantilever. Further practical demands on the production process are the ability for a precise positioning of the spacer element on the nodal line to avoid an additional perpendicular motion of the tip and a reasonable production time. Carbon pillars created by FIB assisted deposition of suitable spatial dimensions fit these requirements well [169] and were therefore used as spacer elements in this dissertation.

The spacer element was placed according to the mark indicating the nodal line and at the center of the cantilever's width to minimize an asymmetry along this dimension of the cantilever which could lead to torsional disturbances of the oscillation. Such a FIB grown carbon pillar spacer element is presented in figure 5.3(4).

Figure 5.3: Bidirectional sensor production steps: (1) commercially available tipless cantilever, (2) removal of the triangular free end, (3) marking the position of the nodal line at a position of $0.217L$ from the free end of the cantilever, (4) deposition of a spacer element by FIB assisted carbon deposition, (5) attachment of the FeCNT interaction tip ((a) direct attachment, (b) using a tungsten needle), (6) finished sensor.

As an experimental detail, dSFM and MFM instruments usually require the cantilever of a SFM sensor to be tilted by a certain angle. This results in the oscillation direction being slightly off-axis with respect to the sample surface. This angle may be taken into consideration when depositing the spacer element, however, to ensure that the tip oscillation in the lateral operation mode is exactly parallel to the sample surface.

5.4.3 Attachment of the interaction tip

After the deposition of the spacer element, any kind of interaction tip could be attached to its free end. The choice of interaction tip depends on the desired force type to be measured. For the experiments in this work, FeCNTs were chosen for magnetic stray field sensitivity to conduct MFM experiments.

For the attachment of the FeCNTs two versions of the standard micromanipulation procedure from section 3.4 were employed as shown in figures 5.3(5a) and (5b). Either the sensor was fixed to the micromanipulator's sample holder to achive a direct attachment of a tube from the FeCNT sample to the spacer element or a FeCNT was first extracted from the FeCNT forest with a tungsten needle and then transferred to the spacer element. The first version has the advantage of simplicity as fewer preparation steps are needed. The advantage of the second variant is given by an increased control on the positioning of the FeCNT on the spacer element and the ability to inspect the FeCNT before the attachment to ensure its suitability for MFM measurements. The latter advantage was a profound one, as there are many different ways an arbitrary FeCNT extracted from the substrate may be unsuitable as a magnetic interaction tip including residual iron particles as well as malformed tubes and fillings [94]. Unfortunately it was impossible to inspect the FeCNTs before extraction from the FeCNT sample. If an unsuitable FeCNT was attached to the spacer element of a bidirectional sensor it had to be removed by an additional micromanipulation procedure, which involved the attachment of a tungsten needle to the FeCNT and ripping it from the sensor. This was possible but included the risk to damage the spacer element and cantilever. While in the beginning of the experiments for this thesis the first variant was employed, in later experiments the second version was chosen.

For MFM experiments it is favorable that the magnetic element of the tip is as close to the sample as possible to increase the possible magnetostatic interaction between tip and sample. If the iron filling of the FeCNT was not extending to the end of the carbon nanotube the unfilled carbon shell was removed by electron beam irradiation in a water atmosphere [90, 91] until only a thin cap remained.

5.4.4 Examination of the completed sensor

The aforementioned steps resulted in completed bidirectional dSFM sensors with FeCNT interaction tips. One such completed sensor is exemplaryly depicted in figure 5.3(6). After the completed fabrication the geometrical properties of the sensors were measured for calibration purposes. To check if the preparation steps resulted in the desired oscillation properties, the sensor was examined by employing a vibration stage in the SEM. A completed bidirectional scanning force microscopy sensor without an angle correction for the spacer element is shown in figure 5.4 along with its two operation modes. The huge amplitudes in these images were excited for demonstration purposes only. In actual dSFM measurements the employed amplitudes are typically in the order of ten nanometers. It has to be noted that any change of the sensor's geometry, as the deposition of the spacer element and the attachment of the FeCNT, changes the exact position of its nodal point. Since the mass and the moments of inertia of the deposited spacer elements and the FeCNTs are much smaller

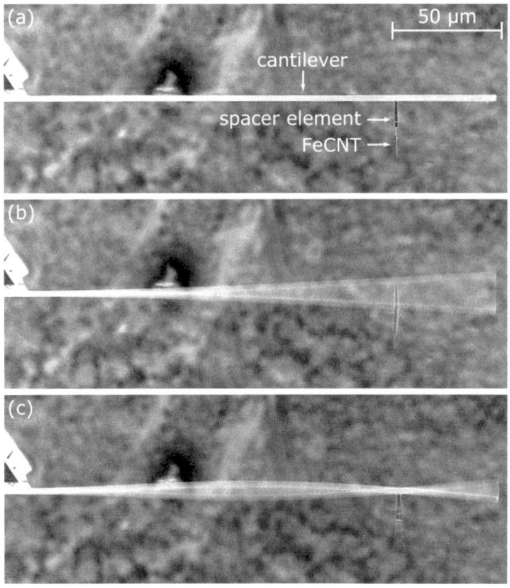

Figure 5.4: Scanning electron microscopy images of a bidirectional sensor: (a)-(c) correspond to the schematic images in figure 5.2.

than those of the cantilever due to their comparatively small dimensions, however, this change was assumed to be small and neglected when creating the sensors. The correctness of this assumption was verified experimentally by studying the mode shape of the second mode of a cantilever as shown in figure 5.4 (c).

5.5 Quantitative experimental validation of the sensor concept

The aim of the following experiments was a verification of the derived formulas for dynamic spring constants and a demonstration of the applicability of the bidirectional sensors for quantitative measurements. As the bidirectional sensors in this work were equipped with iron-filled carbon nanotubes as mentioned above, this study was based on magnetostatic forces between the iron nanowire of the FeCNT and a ferromagnetic test sample.

5.5.1 Bidirectional sensor

In order to conduct the experiment, a bidirectional sensor was prepared from a commercially available doped silicon cantilever[1] according to the fabrication steps described in the previous section. Since the employed dSFM equipment required the sensor to be introduced at an angle of $\gamma = 10°$, this was taken into account during the fabrication of the spacer element. The FeCNT was attached directly to the spacer element and subsequently the unfilled end of the FeCNT was removed by electron beam assisted etching in a water atmosphere.

After the production of the sensor, various SEM images of its structure were taken in order to derive its geometrical properties. This included high magnification images of the free end of the FeCNT to determine the thickness of its carbon cap t_{cap} and the diameter of the contained iron wire d_{Fe}. Corresponding images are shown in figure 5.5.

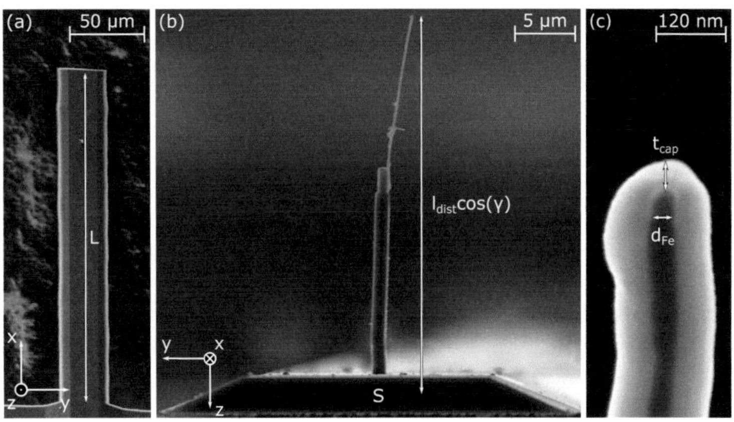

Figure 5.5: Bidirectional sensor equipped with an iron-filled carbon nanotube tip used for the verification of the bidirectional sensor concept: (a) top view showing the cantilever length L, (b) front view of the cantilever's free end: the cantilever cross section area S as well as the projection of the effective spacer length $l_{dist}\cos(\gamma)$ onto the SEM image plane is indicated, (c) free end of the iron-filled carbon nanotube. The arrows indicate the carbon cap thickness t_{cap} and the Fe nanowire diameter d_{Fe}.

Then the sensor was transferred to the dSFM equipment where its resonance frequencies were determined. Furthermore, the calculation of the sensor's dynamic spring constants required the static spring constant of the cantilever. It was obtained according to the

[1]NANOSENSORS TL-FM-10 by NanoWorld AG

Cantilever length	$L = (217.8 \pm 6.5)$ µm
Cross sectional area	$S = (75.9 \pm 4.5)$ µm^2
Effective spacer length	$l_{dist} = (30.1 \pm 0.9)$ µm
Diameter iron filling	$d_{Fe} = (32 \pm 5)$ nm
Thickness of the carbon cap	$t_{cap} = (40 \pm 5)$ nm
Resonance frequency first flex. mode	$f^*_{0,1} = (79.39 \pm 0.01)$ kHz
Resonance frequency second flex. mode	$f^*_{0,2} = (497.64 \pm 0.01)$ kHz
Perpendicular dynamic spring constant	$k^{dyn}_{x_{tip},1} = (4.85 \pm 0.46)$N/m
Lateral dynamic spring constant	$k^{dyn}_{\Delta x,2} = (282 \pm 60)$N/m

Table 5.1: Relevant geometrical and oscillation properties of the bidirectional sensor used to verify the bidirectional sensor concept.

expression given in equation 2.37 with the standard resonance frequency $f^*_{0,1}$ included instead of the angular eigenfrequency:

$$k_L^{stat} \approx \frac{12\pi^2 (f^*_{0,1})^2 \rho S L}{(1.875)^4}. \tag{5.10}$$

The only parameter in this equation that could not be obtained by direct measurements within either the SEM or the dSFM equipment was the mass density, so a literature value for silicon $\rho^{Si} = 2329$ kg/m^3 [170] was used. Based on this calculated static spring constant, the dynamic spring constants were obtained according to the considerations leading to equations 5.3 and 5.7. The obtained properties of the sensor along with the calculated dynamic spring constants for the two operation modes are summarized in table 5.1. The length measurements of the cantilever were estimated to exhibit a maximal error of $\leq 3\%$. This includes the error of the SEM instrument as well as a possible misalignment of the cantilever with respect to the SEM image plane [171, 172]. The obtained maximal errors on the spring constants are comparable to those of similar spring constant determinations found in literature [171, 172].

The uncertainties of the spring constants may also be somewhat underestimated, as some sources of errors were difficult to quantify, for example variations of cross sectional area of the cantilever along its length.

5.5.2 Ferromagnetic reference sample

The objective of the experiments was a verification of the bidirectional sensor's quantitative force sensing capabilities. Therefore, a well defined force field was needed. The iron filling of the FeCNT interaction tips results in a sensitivity of the sensor to magnetostatic interactions, hence a ferromagnetic test sample that provides a well defined magnetic stray field was desirable.

Cobalt-platinum thin film multilayer structures are well known for their strong perpendicular anisotropy for more than two decades [173]. Their magnetic properties result from a complex interplay of many magnetic energy contributions [174]. These contributions include coupling of the Co-layers by the RKKY-interaction[2] mediated by the conduction electrons of the platinum [174] as well as interfacial effects [178] and a polarization of the Pt-layers [174]. The magnitudes of these effects have been found to depend on the thickness of the layers [174, 178] and the general shape of the hysteresis loop of such Co-Pt multilayer structures is influenced by the number of multilayers in the sample [174]. This dependence of the magnetic properties on the layer architecture of Co-Pt thin film multilayer structures makes it possible to optimize a respective sample for specific requirements.

Multilayers with the stacking architecture Pt(5 nm)/[Pt(0.9 nm)/Co(0.4 nm)]$_{100}$/Pt(2 nm) have been introduced as reference sample for MFM tip calibration by Silvia Vock during her PhD work [179]. The layers in such samples are ferromagnetically coupled and the whole layer structure behaves as a single ferromagnet [174]. The stacking structure has been optimized for a strong perpendicular alignment of the magnetization of the sample within each domain and an approximately equal distribution of the domains in the field free case even after being being exposed to a high external magnetic field. This results in a characteristic so called 'constricted' shape of the hysteresis curve with low remanence and coercivity as has been reported for similar Co/Pt multilayers [180]. In the field-free case, these samples show a labyrinth-like band domain structure, which is usually not influenced by typical stray fields emanating from a MFM tip. The sample has been characterized by vibrating sample magnetometry measurements that revealed a saturation magnetization of $M_S^{Co/Pt} \approx 554$ kA/m, a uniaxial perpendicular anisotropy constant of $K_u \approx 517$ kJ/m^3 and a stray field energy constant of $K_d = \mu_0 M_S^{Co/Pt}/2 \approx 192$ kJ/m^3. From these results a dimensionless reduced perpendicular magnetic anisotropy factor of the film of $Q_{mag} = K_u/K_d \approx 2.7$ [181] can be deduced. This value is in accordance with perpendicular magnetization [182, p.499]. Furthermore, these results point to a domain wall width of $\delta_{dw} \approx 24$ nm [183].

[2]Named after Malvin A. Ruderman, Charles Kittel, Tadao Kasuya and Kei Yosida who first described this interaction [175, 176, 177]

5.5.3 Magnetostatic force gradient measurements

With one of these Co/Pt multilayer thin film samples[3] the prepared bidirectional sensor was used to conduct frequency modulation dSFM measurements to probe the magnetostatic forces between the FeCNT tip and the multilayer sample in both operation modes. The experiments were conducted in high vacuum at room temperature. Furthermore, the oscillation amplitude of the first flexural mode was calibrated according to the method described in section 2.6.2.

The measured deflection of a cantilever is proportional to the slope of the cantilever at its free end [77]. This was used to derive the amplitude of the oscillation in the lateral operation mode of the sensor. To this end, the measured amplitude signal was recalculated into the corresponding slope of the cantilever at its free end by using the mode shape of the cantilever in the first flexural oscillation mode $u_1(x)$ and its derivative $u_1'(x)$ as given by equations 2.23 and 2.24. Then, the free constant in the derivative $u_2'(x)$ of the mode shape of the second flexural mode from equation 2.24 that corresponded to this slope at the free end of the cantilever was derived numerically. This constant was then employed to calculate the cantilever slope at the nodal point $u_2'(x_0)$. The oscillation amplitude of the FeCNT's free end was then obtained by using equation 5.4.

With the parameters used in the experiment the tip oscillation amplitudes were found to be ≈ 10 nm in the perpendicular sensitivity mode and ≈ 9 nm in the lateral sensitivity mode. According to equation 2.11 the frequency shift signal does not depend on the amplitude, and therefore the small difference of ≈ 1 nm is neglected.

To exclude the influence of the sample topography all measurements were performed at a nominal mean tip-sample distance of $z_{scan} = (80 \pm 5)$ nm as obtained according to the method from section 2.6.3 with the corresponding measurements in the perpendicular operation mode. Furthermore, the bias potential was adjusted to compensate electrostatic force contributions as discussed in section 2.5.3. The scan range was chosen to cover a 5 μm $\times$ 5 μm area of the same sample position with a resolution of 1024×1024 data points in both modes at a measurement speed of 1 line/s.

The data obtained by these measurements are shown in figures 5.6(a) and (b)[4] Here, the frequency shift signal provided by the dSFM instrumentation was already recalculated into force gradient data by using the obtained dynamic spring constants from table 5.1 together with the corresponding equations 5.8 and 5.9. The data was corrected for thermal

[3] Provided by Denys Makarov and Cornelia Krien (IFW Dresden), Sample-No.:d3460

[4] Please note that the sign of the force gradient data in the corresponding publication [184] was reversed due to an error in the evaluation software. It is emphasized that this does not change the general results, however. The force gradient data shown in figures 5.6, 5.7 and 5.10 in this thesis were calculated with the correct sign.

drift of the sensor's resonance frequency during the measurements by a plane subtraction. Furthermore, small thermal drifts of the sample stage were corrected by extraction of overlapping image parts from the measurements. The software WSxM [185] was used to conduct these corrections. Due to the perpendicular magnetization of the domains, the perpendicular sensitivity measurement in figure 5.6(a) also illustrates the band domain structure of the multilayer sample [182]. The lateral measurement data shows a low signal at areas, where the oscillation direction of the FeCNT was oriented parallel to nearby domain boundaries and a high signal at areas where the domain boundaries and the measurement direction intersected at a angle of roughly 90°. This behavior is expected for a lateral magnetic force gradient measurement and serves as a first affirmation of the lateral sensitivity of the bidirectional sensor.

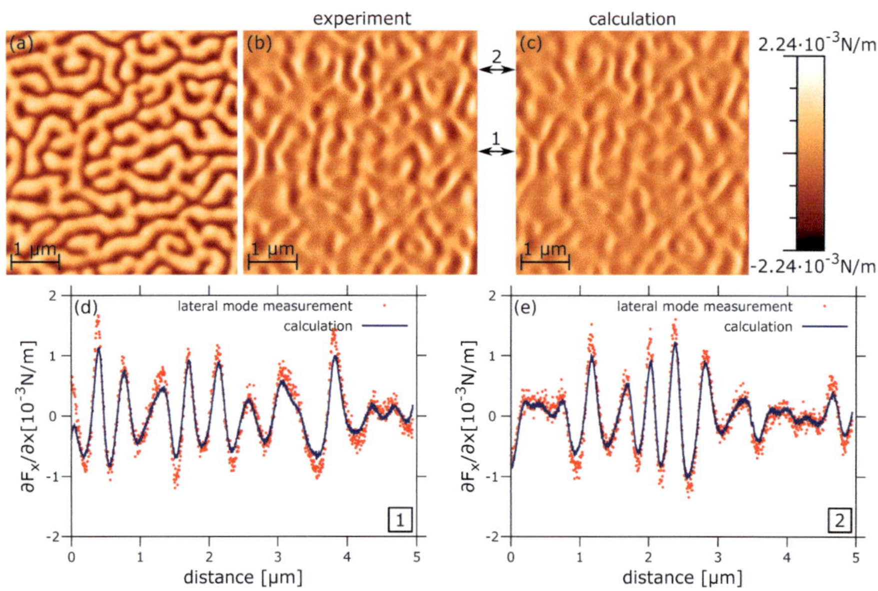

Figure 5.6: Magnetostatic bidirectional force gradient data and corresponding line sections: (a) experimental perpendicular force gradient map, (b) experimental lateral force gradient map, (c) lateral force gradient map calculated from (a), (d) and (e) line sections of the force gradient data shown in (b) and (c) at the positions that are indicated by arrows and numbers.

5.5.4 Calculation of lateral magnetostatic force gradients

Two-dimensional Fourier transforms are often used in the calculations regarding complex MFM measurements to increase the computational speed. Such Fourier transforms replace the lateral coordinates in real space (x, y) with the respective spatial frequencies (c_x, c_y)[5]. As a result of this transformation, the calculation of the spatial derivatives reduces to a simple multiplication operation. Assuming a purely magnetostatic measurement of a source-free stray field, it is possible to derive the whole magnetic stray field of a sample $\widetilde{\vec{H}}(c_x, c_y, z)$[6] in this Fourier space from the knowledge of its perpendicular component $\widetilde{H}_z(c_x, c_y, z)$ [186]:

$$\widetilde{\vec{H}}(c_x, c_y, z) = -\frac{\widetilde{\vec{\nabla}}}{c}\widetilde{H}_z(c_x, c_y, z) \tag{5.11}$$

with $\widetilde{\vec{\nabla}} = (ic_x, ic_y, -c)$ being the Fourier space nabla-operator for this case and $c = \sqrt{c_x^2 + c_y^2}$ denoting the absolute value of the in-plane wave vector. The only exception is the point $(c_x = 0, c_y = 0)$, which corresponds to the mean value of the stray field [179]. From equation 5.11 it can be derived by using the components of the nabla-operator that

$$\widetilde{\frac{\partial H_x}{\partial x}} = -\frac{c_x^2}{c^2}\widetilde{\frac{\partial H_z}{\partial z}} \tag{5.12}$$

holds. In a MFM experiment this magnetic field gradient is translated to a force gradient by the magnetic properties of the tip. In the case of a magnetic tip with a monopole-like characteristic, like the FeCNTs used for the sensors in this chapter, both sides of equation 5.12 can simply be multiplied by $-q\mu_0$ leading to [179]:

$$\widetilde{\frac{\partial F_x}{\partial x}} = -\frac{c_x^2}{c^2}\widetilde{\frac{\partial F_z}{\partial z}}. \tag{5.13}$$

A monopole-like tip characteristic is not required for the validity of this equation, however, and it can be used for any MFM tip [179].

For the calculation of the lateral force gradients experimentally obtained perpendicular force gradient data maps need to be transformed into Fourier space. Then the lateral force gradients can be calculated according to equation 5.13. The real space lateral force gradient maps can be acquired by applying an inverse Fourier transformation. A software package provided by Silvia Vock (IFW Dresden) that implements the required calculations was employed in this work.

[5] The common notation of the spatial frequencies in Fourier transforms $\vec{k}$ has been replaced by $\vec{c}$ in order to avoid confusion with the spring constants

[6] The tilde sign (' ˜ ') denotes terms in Fourier space

5.5.5 Comparison between measurement and calculation

The experimental perpendicular force gradient data depicted in figure 5.6(a) was used together with the method and the corresponding software package described in the last section to obtain the calculated lateral force gradient data shown in 5.6(c). A comparison between these calculated results and the corresponding experimental lateral data (see figure 5.6(b)) showed that the general shapes of the force gradient maps are very similar. Hence, the second flexural mode of the sensor indeed provides the claimed lateral sensitivity. A more detailed quantitative comparison was conducted by extracting line sections of the data. Two exemplary line sections are shown in figures 5.6(d) and (e). As can be seen in these line sections the measured data exhibits a slightly higher signal than the calculated one.

For an estimation of the errors of these data sets, several factors had to be taken into account. First of all, the calculated data was based on experimental frequency shift data recalculated according to equation 5.8 into force gradient data by using a factor of $-2k^{dyn}_{x_{tip},1}/f^*_{0,1} = (-0.122 \pm 0.012)\ 10^{-3}\text{Ns/m}$ with an error on the respective spring constant and resonance frequency as given in table 5.1. This factor does not depend on the spatial coordinates or the spatial frequencies. This together with linearity of equation 5.13 and the Fourier transformation results in the relative error of this factor to be directly translated to a relative error of the calculated data. For the experimental data two sources of errors were taken into account: the maximal error from the recalculation factor $-2k^{dyn}_{\Delta x,2}/f^*_{0,2} = (-1.132 \pm 0.240)\ 10^{-3}\text{Ns/m}$ from equation 5.8 and the margins of the signal noise level of the instrument, which corresponds to typically ± 120 mHz for the chosen experimental parameters and the settings for the phase-locked loop[7]. The results from these considerations are depicted in figure 5.7. The data already shown in figure 5.6(e) was used and only every 5th experimental data point was evaluated for better visibility but to following discussion is valid for all data points. Some, although very few isolated data points are not in accordance with the calculation within the margins of error of the measurement. The deviation of most of these are minimal, however, and can be explained by unaccounted errors induced by the phase-locked loop in response to quickly changing signals. A slower measurement speed could help to solve this but also leads to an increased thermal drift during the duration of the measurement. As this introduces a corresponding difficult to quantify error into the experiment, this was avoided in the experiment. Two experimental data points that do not match the calculated data in the line section, one of which is visible in 5.7 at a section distance of ≈ 4.23 µm, however, cannot be explained by

[7]Proportional gain 0.75 Hz/deg, integration time 4 ms

such an influence of the PLL. The can only be considered true faults in the measurement caused by unknown instabilities of the system or external influences on the equipment. Please note that very close to the edges of the force gradient data maps the experimental and the calculated data are sometimes not in agreement. This can be attributed to an artifact from the calculation in Fourier space induced by using an finite perpendicular data map.

Nevertheless, the vast majority of the measured data points is within the margins of error in quantitative agreement with the calculation, thus substantiating the correctness of the ratio between the calculated spring constants.

Since already the force gradient data for the perpendicular direction was calculated based on the calculated perpendicular operation mode spring constant, this data could not be used to verify the absolute force gradients and an independent experimental validation is needed. Such a validation is presented in the next section.

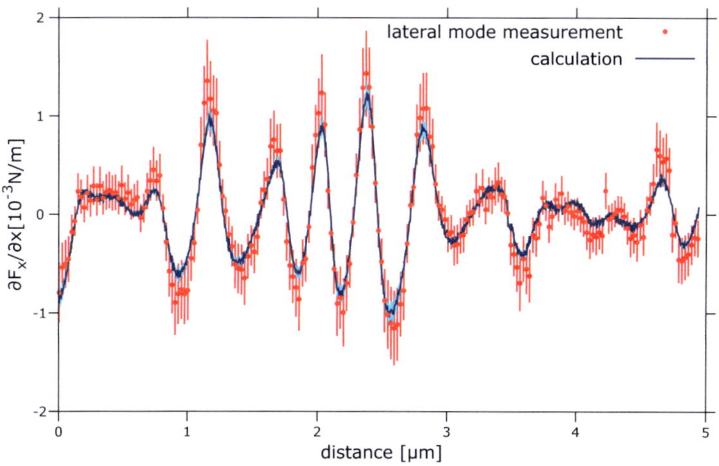

Figure 5.7: Error assessment of the bidirectional force gradient data shown in figure 5.6 (e). For better visibility only every 5th experimental data point is shown. The light blue area indicates the error of the calculated data.

5.5.6 Determination of the magnetic moment of the tip

The verification of the absolute force gradient was provided by the possibility to characterize the magnetic behavior of both the FeCNT tip and the well-known multilayer sample by independent means. In Fourier space it can be shown that the Fourier transforms of the perpendicular force gradients and the tip's perpendicular stray field gradient are connected by the Fourier transform of the sample's effective surface charge pattern $\tilde{\sigma}^*(c_x, c_y)$ via [187]:

$$\widetilde{\frac{\partial F_z}{\partial z}} = \left(\widetilde{\frac{\partial H_z}{\partial z}}\right)_{tip} \cdot \tilde{\sigma}^*(c_x, c_y). \tag{5.14}$$

For the employed Co/Pt multilayer the effective surface charge pattern can be deduced from a qualitative perpendicular sensitivity MFM measurement without the need to know the spring constant of the cantilever, if saturation magnetization, film thickness and domain wall width of the sample are known [187]. In this context it has to be mentioned that small deviations of the magnetization from the perpendicular direction are expected for a reduced perpendicular anisotropy coefficient of $Q_{mag} = 2.7$. It has been shown that these small deviations can be considered by a reduced saturation magnetization $M_{S,red}^{Co/Pt} = \xi M_S^{Co/Pt}$ with $\xi = 2/(1 + \sqrt{1 + 1/Q_{mag}})$ when the effective surface charge pattern is calculated [132]. Applying a deconvolution method involving a Wiener invert filter and the effective surface charge pattern as described in the work of Silvia Vock [179] to experimentally obtained perpendicular force gradient data, results in the stray field gradient of the tip at the sample surface for the same mean magnetic tip-sample distance as the one the experimental data was obtained at.

The magnetic field gradient of the FeCNT tip was calculated according to this method[8]. A corresponding line section through symmetry center of the calculated data is shown in figure 5.8.

Because of the monopole-like stray field characteristic of the FeCNT tips, the calculated stray field gradient can be fitted by that of a monopole [179]. With the the magnetic monopole moment q_{TF} and the distance r_{TF} as the free parameters [188, p.28]:

$$\frac{\partial H_z}{\partial z}(x) = \frac{q_{TF}}{4\pi} \frac{x^2 - 2r_{TF}^2}{(x^2 + r_{TF}^2)^{5/2}}. \tag{5.15}$$

The resulting standard errors of the fit parameters as derived by the fit software, neglect the error on the input parameters of the calculation software. The calculations implemented by the software are complex, thus direct derivation of the errors on the results was not possible. An estimation of the errors of the tip field derived from such a calculation, although for a different dSFM equipment, yielded a relative error of up to 15% [179, p.46] which was

[8]The calculation software and required multilayer sample parameters were provided by Silvia Vock (IFW Dresden).

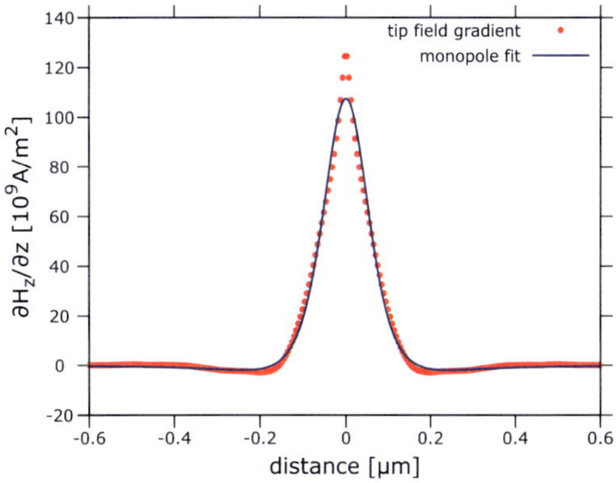

Figure 5.8: Linesection of the stray field gradient of the iron-filled carbon nanotube tip. The mean magnetic tip-sample distance for this stray field gradient corresponds to the measurement presented in figure 5.6(a) from which the data was derived. A monopole stray field gradient fit according to equation 5.15 was applied to the data. The resulting values of the fit parameters were found to be $q_{TF} = (-1.18 \pm 0.03) \cdot 10^{-9}$ Am, $r_{TF} = (121 \pm 1)$ nm. (The stray field gradient was calculated with the software and sample parameters provided by Silvia Vock (IFW Dresden). The data was fitted with the software 'Origin 8.6' and the given errors are the standard errors calculated by the software.)

taken into consideration for an estimation of a maximum errors of the fit results. With the thus obtained margins of error, the fit resulted in the parameters

$$q_{TF} = (-1.18 \pm 0.27) \cdot 10^{-9} \text{ Am}; \quad r_{TF} = (121 \pm 1) \text{ nm}. \tag{5.16}$$

The small deviations between the monopole fit and tip stray field especially at small distances from the symmetry center that are visible in figure 5.8 may be explained by the non-vanishing lateral dimensions of the iron core. Please note that the mean magnetic tip-sample distance is different from the mechanically obtained mean tip-sample distance z_{scan} due to the carbon shell protecting the magnetic nanowire. Please note, that the negative sign of the monopole-like moment is based on the sign convention in the calculation software only.

The geometrical distance r_{geom} of the iron wire's apex to the sample surface is given by the sum of the mean tip-sample distance z_{scan} as measured in the dSFM equipment and the thickness of the carbon cap t_{cap}. The absolute value of the monopole-like moment $|q_{geom}|$ of the iron filling of a FeCNT can be obtained by multiplying the saturation magnetization M_S^{Fe} of iron with the cross sectional area $\pi d_{Fe}^2/4$ of the iron filling [132].

With the geometrical parameters, which were derived from SEM images, as listed in table 5.1 the resulting parameters are found to be:

$$|q_{geom}| = M_S^{Fe} \pi \frac{d_{Fe}^2}{4} = (1.4 \pm 0.4) \cdot 10^{-9}\ \text{Am}; \quad r_{geom} = z_{scan} + t_{cap} = (120 \pm 10)\ \text{nm}. \quad (5.17)$$

The relatively large error interval of the geometrical monopole moment was caused by the derivation of the diameter of the iron filling from SEM images.

The magnitudes of the effective monopole parameters derived according to both methods are in good agreement. As both parameter sets were obtained by independent measurements this validates the absolute force gradient calculation and thus the applicability of the bidirectional sensors for quantitative force gradient measurements. Furthermore, the very good agreement of the geometrical distance r_{geom} to the fitted magnetic tip-sample distance indicates that the method to determine the tip-sample distance as introduced in section 2.6.3 yields satisfactory results within the given margins of error.

With the knowledge of the monopole-like moment of the FeCNT, it is possible to recalculate both the lateral and the perpendicular frequency shift into a magnetic stray field gradient by applying equation 2.45 for a monopole point probe and using the spring constants in accordance with equations 5.8 and 5.9. It has be noted, however, that both methods only reliably provide the absolute value of the effective monopole moment of the FeCNT tips. The real sign of the effective monopole moment can only be obtained by orienting the magnetization of the iron filling of the FeCNT by applying a magnetic field with a well known direction prior to the measurement. This was not necessary for these verification experiments, however.

All of the above considerations are only valid if both the spacer element and the carbon nanotube are rigid in a sense that their resonance frequencies far exceed that of the cantilever. After conducting the previous measurements, this was verified by employing a vibration stage in the SEM. The resulting images can be found in figure 5.9 and no trace of non-rigid behavior, i.e. a bending of spacer element or FeCNT, in the lateral operation mode was found. The FeCNT exhibited a fundamental resonance frequency of $f_{CNT}^* \approx 1.08$ MHz and the resonance frequency of the carbon pillar should even be larger [169] than the much thinner carbon nanotube. Thereby, the approximately rigid behavior of the spacer nanotube combination was confirmed. As will be shown later in chapters 6 and 7, however, even at such large separations of the resonance frequencies and with the absence of visual bending in the SEM images, very small deviations of the nanotube from rigid behavior are possible and influence the sensors lateral frequency shift signal. This can result in a slight signal amplification, which might have contributed to the experimental force gradients being somewhat larger than the calculation in figure 5.6, especially for the extremal values. However, the larger force gradients could also have been caused by the influence of the systematic errors of the geometric measurements when calculating the effective spring constant, thus hiding such a small amplification.

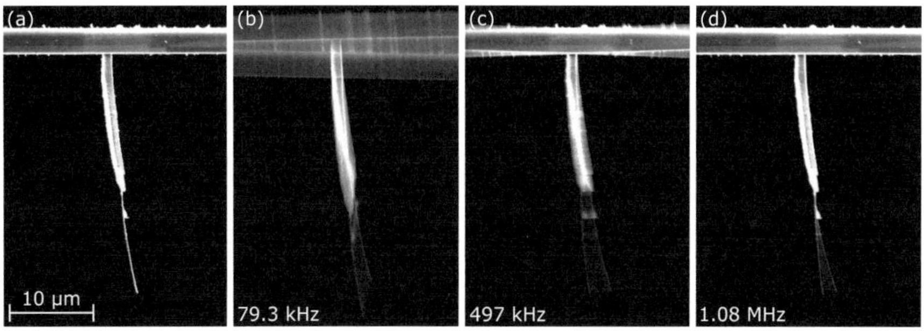

Figure 5.9: Vibration study of the bidirectional sensor used for the experimental verification of the quantitative bidirectional measurements: (a) no excitation, (b) first flexural mode of the cantilever for perpendicular sensitivity, (c) second flexural mode of the cantilever for lateral sensitivity, (d) first flexural mode of the FeCNT interaction tip. The approximate excitation frequencies are indicated.

5.6 Extension to tridirectional scanning force microscopy

As it was mentioned in the introduction of this chapter in section 5.1.2, the torsional resonance mode LFM technique is compatible with standard dSFM sensors. As bidirectional sensors can be considered standard dSFM sensors with a specially positioned interaction tip they should also be suitable for a torsional resonance approach. As the bidirectional sensors and the torsional resonance mode implement a movement in two different lateral directions, a combination of both concepts allows a measurement of force gradients along all three spatial directions.

5.6.1 Torsional resonance lateral force microscopy

In order to conduct quantitative LFM based on the excitation of the first torsional resonance mode, the corresponding spring constants for this mode have to be derived. As this is an already established method the spring constants can be obtained according to formulas available in literature, so only a short overview is presented here.

The calculation of the torsional spring constant is based on the differential equation [189]

$$\frac{GK}{L^2}\frac{\partial^2\psi(\xi,t)}{\partial\xi^2} - \rho I_p \frac{\partial^2\psi(\xi,t)}{\partial t^2} = T(\xi,t) \tag{5.18}$$

for the angulat deflection $\psi(\xi,t)$ about the main axis of a cantilever at a position $\xi = x/L$ along its length. In this equation GK is the torsional rigidity, I_p the polar moment of the cross section with respect to the torsional axis, ρ the mass density and $T(\xi,t)$ an applied torque per unit length. For a cantilever beam the mode shape of the first torsional mode is then given by [190]:

$$\psi(\xi) = C_{tors}\sin(\frac{\pi}{2}\xi) \tag{5.19}$$

with the constant C_{tors}. The static torsional spring constant for an angle deflection Φ of a cantilever with respect to its torsion axis and its free end can be written as [189]:

$$k^{stat}_{L,\Phi} = \frac{GK}{L}. \tag{5.20}$$

The first torsional resonance frequency of the beam [189]

$$\omega^2_{\Phi,1} = \frac{\pi^2}{4}\frac{GK}{\rho I_p L^2} \tag{5.21}$$

can be used to eliminate the torsional rigidity GK in analogy to the elimination of the flexural rigidity EI in the normal static spring constant of the cantilever in section 2.4. The torsional spring constant can then be used to calculate the corresponding static lateral spring constant

$$k^{stat}_{L,t} = \frac{k^{stat}_{L,\Phi}}{l^2_{dist}} = \frac{4\omega^2_{\Phi,1}\rho I_p L}{\pi^2 l^2_{dist}} \tag{5.22}$$

by geometrical considerations involving the effective spacer length l_{dist} [191]. This static spring constant can be recalculated into a dynamic spring constant $k^{dyn}_{L,t}$ by including a factor of $\pi^2/8$ [192]. These equations are valid for standard dSFM sensors. For the bidirectional sensors, however, the position $x_{tip} \approx 0.783L$ of the spacer element carrying the interaction tip has to be considered. The expression for the amplitude of the oscillation of the tip's apex can be derived again by simple trigonometry

$$\Delta y = l_{dist}\sin(\psi(x_{tip}/L)) \approx l_{dist}\psi(x_{tip}/L). \tag{5.23}$$

The same energy considerations as were employed to derive equation 5.6 for the second flexural mode of the bidirectional sensor can then used to obtain a relation for the dynamic lateral spring constant with respect to the interaction tip:

$$k_{\Delta y,t}^{dyn} = \frac{\sin^2(\frac{\pi}{2})}{\sin^2(\frac{\pi}{2}\frac{x_{tip}}{L})} k_{L,t}^{dyn} \approx \frac{1}{\sin^2(0.783\frac{\pi}{2})} k_{L,t}^{dyn} \approx 1.13\frac{\pi^2}{8} k_{L,t}^{stat}. \tag{5.24}$$

Combining all these considerations results in the dynamic lateral spring constant for the first torsional mode of the cantilever with the interaction tip as a point of reference:

$$k_{\Delta y,t}^{dyn} = \frac{\omega_{\Phi,1}^2 \rho I_p L}{2 l_{dist}^2 \sin^2(\frac{\pi}{2}\frac{x_{tip}}{L})}. \tag{5.25}$$

To apply this equation for the cantilevers that were used in these experiments, the polar moment of area

$$I_p = \frac{h(16h^2ab + 4h^2b^2 + 4h^2a^2 + 3a^4 + 6a^2b^2 + 6a^3b + 6ab^3 + 3b^4)}{144(a+b)} \tag{5.26}$$

for an isosceles trapezoid with respect to the main axis of the torsion is required. This can be derived from the perpendicular axis theorem [28, p.123] using the literature values for the second moments of area [193, p.E80]. Here, h is the height, a the upper and b the lower length of the trapezoid. Finally, frequency shift data obtained with the first torsional mode can be related to the lateral force gradient by:

$$\Delta f_y = -\frac{f_{\Phi,1}}{2k_{\Delta y,t}^{dyn}}\frac{\partial F_y}{\partial y} \left(\approx -\frac{f_{\Phi,1}^*}{2k_{\Delta y,t}^{dyn}}\frac{\partial F_y}{\partial y} \right). \tag{5.27}$$

5.6.2 Tridirectional force gradient measurement

For the experimental demonstration of compatibility of the bidirectional sensors with the torsional resonance mode a new sensor was fabricated, this time a tungsten needle was employed to attach the FeCNT. The properties of this sensor are listed in table 5.2. As the resonance frequency of the first torsional mode was much higher than that of the second flexural mode, a very short FeCNT was used as a magnetic interaction tip to ensure a rigid behavior of the FeCNT. This resulted in a short effective spacer length l_{dist} and a corresponding high effective spring constant in the lateral operation mode based on the flexural vibration.

With this sensor dSFM measurements were conducted on the same ferromagnetic test sample as used above. The mean tip-sample distance was chosen to be $z_{scan} = (65 \pm 5)$nm

Cantilever length	$L = (215.0 \pm 6.5)$ µm
Cross sectional area	$S = (74.5 \pm 4.5)$ µm^2
Effective spacer length	$l_{dist} = (13.87 \pm 0.2)$ µm
Resonance frequency first flex. mode	$f^*_{0,1} = (79.98 \pm 0.01)$ kHz
Resonance frequency second flex. mode	$f^*_{0,2} = (503.60 \pm 0.01)$ kHz
Resonance frequency first tors. mode	$f^*_{\Phi,1} = (1002.9 \pm 0.1)$ kHz
Perpendicular dynamic spring constant	$k^{dyn}_{x_{tip},1} = (4.77 \pm 0.45)$ N/m
Lateral dynamic flex. spring constant	$k^{dyn}_{\Delta x,2} = (1.27 \pm 0.23)$ kN/m
Lateral dynamic tors. spring constant	$k^{dyn}_{\Delta y(x_{tip}),t} = (325 \pm 68)$ N/m

Table 5.2: Relevant properties of the bidirectional sensor used to verify the compatibility with quatitative torsional resonance mode measurements.

and the measurement resolution was reduced to 512 × 512 points as this was deemed sufficient to depict the rather broad stray field features of the sample. The remaining experimental procedure was almost identical to the measurements discussed in the previous sections of this chapter.

Differences arose from the more difficult measurement conditions as discussed in the following: the high effective spring constant of the second flexural based lateral operation mode resulted in an tiny frequency shift signal below the noise levels of the previous measurement conditions. Therefore, a very low measurement speed of (1/3.5) line/s combined with a change of the parameters of the phase-locked-loop[9] had to be applied to decrease the noise levels. However, this resulted in an increased overall measurement time and a notable thermal drift of the mean-tip sample separation during the measurement. As mentioned before this introduces a unquantifiable error on the experimentally obtained data. Even then, with the low measurement speed and adjusted phase-locked-loop parameters the maximal frequency shift signals in the order of ±100 mHz were only slightly larger than the noise levels. Furthermore, due to the short effective spacer length an oscillation amplitude of the interaction tip of 2 nm was used for the experiment.

A different matter was the lateral measurement based on the torsional resonance mode as the dSFM equipment does not support this measurement mode. Nevertheless, it was possible to use the lateral detector signal, which was designed for static lateral measurements, to detect the torsional vibrations. Still, a dedicated method to reliably excite the torsional vibrations of the cantilever is not available in the instrument. It has been shown, however, that a vertical mechanical excitation usually used to excite flexural vibrations of cantilevers can also excite the torsional vibrations of the sensor's cantilever [156] although with a very

[9]Proportional gain 0.15 Hz/deg, integration time 4.5 ms

low efficiency. Setting the mechanical excitation to its maximal amplitude resulted in a detectable although very low amplitude signal from the torsional vibration mode. While this low signal could still be used to implement a torsional resonance mode frequency modulation measurement, it introduced considerable noise to the data. To compensate this to a certain extend a lower measurement speed of (1/1.8) line/s with adjusted phase-locked-loop parameters[10] was chosen.

5.6.3 Evaluation of the tridirectional data

The experimental frequency shifts were again corrected for thermal drift and recalculated into force gradients. The results are shown along with corresponding calculated data in figure 5.10. By comparing the general shape of these force gradient maps a good qualitative agreement between the measured data and the calculations was found in both lateral data acquisition modes.

These results verify the compatibility of the bidirectional sensors with the torsional resonance mode to implement tridirectional dynamic mode scanning force microscopy. The high noise levels of the measurements and the unknown error induced by the thermal drift of the sensor position during the longer measurement process make a reasonable quantitative discussion of the results impossible. Nevertheless, the good agreement between the measured and the calculated torsional resonance mode data is a strong indication that even quantitative tridirectional data evaluation is possible. A future experimental verification of this could be obtained by an experiment using a bidirectional sensor with an optimized longer spacer element to increase the lateral spring constants as well as dSFM equipment with a dedicated torsional resonance mode excitation and detection scheme.

Please note that, according to Maxwell's equations, for magnetic fields [64, p.3]

$$\vec{\nabla} \cdot \vec{B} = \frac{\partial B_x}{\partial x} + \frac{\partial B_y}{\partial y} + \frac{\partial B_z}{\partial z} = 0 \tag{5.28}$$

holds. Hence, if two field gradient components are known the third can easily be calculated. This means that tridirectional dSFM measurements are not required for magnetostatic forces. Nevertheless, these experiments suggest the applicability of quantitative tridirectional dSFM measurements to any kind of force gradients, including those without such an easy way to calculate the third component.

[10]Proportional gain 0.3 Hz/deg, integration time 4 ms

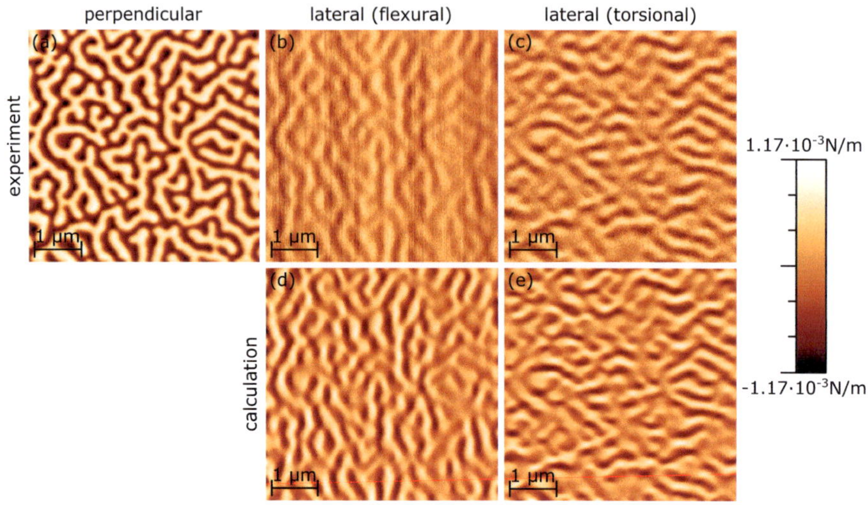

Figure 5.10: Magnetostatic force gradient data of a tridirectional measurement: (a) experimental perpendicular force gradient map, (b) experimental lateral force gradient map obtained with the lateral operation mode based on the cantilever's second flexural vibration, (c) experimental lateral force gradient map obtained with the torsional resonance mode, (d) and (e) lateral force gradient maps calculated from (a) corresponding to (b) and (c) respectively.

5.7 Advantages and challenges of the bidirectional sensor concept

In this chapter the viability of the bidirectional sensor concept for quantitative force gradient measurements was demonstrated. It is emphasized that despite magnetostatic forces were used for the verification of the quantitative operating principle of the sensor, both the sensor concept and calculation of dynamic spring constants were derived based on mechanical considerations only. Therefore, these considerations are valid for any type of force and the bidirectional sensor concept is applicable to general dSFM.

As there are well established methods of lateral force microscopy available, it is important to discuss similarities and differences of the methods. As opposed to the pendulum geometry the bidirectional sensors provide both perpendicular and lateral sensitivity with an easy in-situ switching between the modes by choosing the respective resonance frequency and are in this aspect similar to torsional resonance mode lateral force microscopy. Additionally, the presented bidirectional sensor concept only relies on vertical excitation and detection schemes and is therefore compatible with any standard dSFM setup, whereas the torsional resonance mode usually requires special equipment. An appertaining advantage is the easy calculation of the lateral oscillation amplitude from the perpendicular one, whereas the torsional mode often requires an additional amplitude calibration as two different deflection directions need to be detected [194]. An additional advantage of the bidirectional sensors is their demonstrated compatibility with the torsional resonance mode to obtain a switchable force gradient sensitivity along all basic spatial directions.

A remaining challenge of the bidirectional sensors is their high effective lateral spring constant which results in a low sensitivity of the lateral mode. In this aspect they perform poorly compared to the pendulum geometry that can employ cantilevers with extremely low spring constants. However, this disadvantage is shared with the torsional resonance mode, which also suffers from high effective spring constants, but not as pronounced as the bidirectional sensors [195, 192].

A way to improve the lateral spring constant is to fabricate longer spacer elements as is obvious from the calculated lateral mode spring constant in equation 5.7. Besides considerations on the practicability of the fabrication of longer spacer elements by the FIB induced deposition of carbon, however, this approach is limited by the resonance frequency of the spacer element. The resonance frequency of the spacer element depends on its length and will be reduced for a longer carbon pillar of otherwise equal dimensions [169]. At a certain length it cannot be considered rigid anymore and this results in an invalidation of the considerations above to implement a quantitative measurement. Of course this loss of rigidity can be counteracted by increasing the other spatial dimensions of the spacer element, especially the one along the cantilever's length, but it has to be kept in mind that a large spacer element has to be taken into account when calculating the mode shape of the cantilever and connected to this its dynamic spring constants. The resulting optimization problem is of importance for practical applications and should be explored in the future.

Nevertheless, by dropping the requirement of rigidity a more elegant solution to increase the lateral signal of the bidirectional sensors is proposed in chapter 6: the co-resonant sensor concept.

6 Co-resonant sensor concept

Iron-filled carbon nanotubes similar to the ones used as magnetic tips for the bidirectional sensors have already been proven to be highly sensitive force sensors themselves when oscillated at their fundamental resonance frequency [6]. Therefore, tapping this sensitivity and exploiting it for dSFM purposes is a very attractive prospect. The detection of their oscillatory state, however, is more difficult as compared to dSFM cantilevers due to their reduced lateral dimensions. In this chapter a method to meet this challenge is proposed. Based on a simple coupled harmonic oscillator model with asymmetric parameters of the subsystems, a concept is derived to combine micro- and nanoscaled mechanical resonators to acquire sensors with increased force gradient or mass sensitivity and an easy to detect oscillation signal [196].

6.1 Oscillating micro- and nanomechanical systems

Oscillating micro- and nanomechanical systems are routinely applied in science and technology. In fact, one of these applications has already been extensively discussed in this work: the microstructured dSFM force sensors. Others include radio frequency filters [197], accelerometers [198], sensitive mass sensors [199, 200] and force sensors for other experimental techniques like magnetic resonance imaging [201] and cantilever magnetometry [133].

6.1.1 Oscillating force- and mass sensors

Micro- and nanoscaled mechanical resonators employed as sensors are commonly based on the same operating principles as the dSFM sensors that have been discussed in chapter 2. The influence of a physical quantitiy to be measured alters the eigenfrequency or the damping of the resonators with corresponding changes to its amplitude and phase with respect to the excitation. These changes can be measured and the desired information about the physical quantity derived from the data. In simple cases, the resonators can be approximated by a simple spring-mass harmonic oscillator with an effective spring constant

k_{eff} and mass m_{eff} as was derived for the dSFM cantilevers in section 2.2.3. Considering the influence of a small mass change Δm in addition to the already discussed external force gradient $\Delta k = -\partial F_s/\partial s$ acting on the oscillator, the resulting eigenfrequency

$$\omega_{\Delta k,\Delta m} = \sqrt{\frac{k_{eff} + \Delta k}{m_{eff} + \Delta m}} \tag{6.1}$$

can be linearized. The corresponding frequency shift is given by

$$\Delta\omega \approx \frac{\omega_0}{2}\left(\frac{\Delta k}{k_{eff}} - \frac{\Delta m}{m_{eff}}\right). \tag{6.2}$$

From this equation the advantages of further miniaturization of mechanical oscillation based sensors can be derived. The frequency shift signal depends on the eigenfrequency and the relative change of the mass and force gradient. A proportional reduction of the spatial dimensions of a cantilever oscillator results in a decrease of both, its mass and spring constant, thus increasing the frequency shift signal produced by a given force gradient or mass change. Furthermore, since the spring constant scales linearly, whereas the mass exhibits a cubic scaling behavior, the eigenfrequency will increase for proportionally reduced dimensions. This will increase both the frequency shift signal and the possible measurement speed [7, p.19]. Hence, in order to create highly sensitive cantilever based sensors it would be favorable to use nanoscaled systems. Typical nanosystems that have already been demonstrated to be viable highly as sensitive force [6] and mass sensors [202] are carbon nanotubes. The determination of their oscillatory state, however, is typically based on more complex detection mechanisms than the simple optical beam deflection method often employed for dSFM sensors [6, 203, 204].

6.1.2 Coupled micro- and nanomechanical sensor systems

Over the last years coupled oscillating micro- and nanomechanical sensor systems have attracted attention due to their potential to increase sensitivity in mass sensing or to solve application-related challenges. In the field of dSFM a system of two coupled cantilevers with very different resonance frequencies has been proposed. By exciting both resonance frequencies of the cantilevers at the same time a dual frequency modulation technique can be implemented to simultaneously measure tip-sample interaction force curves and topography [205].

Many coupled micromechanical oscillator systems have been studied with applications for mass sensing. One approach uses a system of two identical coupled cantilever beams that employ the principle of mode localization. These mode localization based sensors exhibit a large change of the oscillation amplitude, which can be used as a measurement signal, if the initial symmetry of an oscillating system consisting of two coupled identical subsystems is broken by an additional mass [206]. Several extensions to his approach have been proposed as the usage of three instead of two identical coupled cantilevers [207] or a new sensing paradigm [208].

Other kinds of coupled micro- and nanomechanical mass sensors employ the coupling as a remote sensing method. It has been demonstrated recently by Biswas et al. that the mass loading of each individual string in an array of coupled similar silicon nitride nanostrings can be deduced from a vibration measurement of one single string of the array [209]. Another application of remote sensing reported by DeMartini et al. involved an array of smaller mass sensing resonators that were read out by detecting the oscillation of a single bigger oscillator coupled to the array [210].

In addition to applied research which aims to create sensor devices, coupled micro- and nanomechanical resonators have also been used as specimen systems to investigate more fundamental aspects of oscillatory behavior. This includes experiments to study nonlinear dynamics [211] as well as parametric [212, 213] and synchronized oscillations [212, 214].

All of these experiments have in common that either the coupled systems are of approximately the same size or their respective resonance frequencies are far apart. Furthermore, there are experiments that demonstrated a coupling between different spatial modes of the same oscillator [215]. Of course, there are technological applications of coupled mechanical systems of different sizes with identical resonance frequencies. An example for these are tuned mass dampers that are used to reduce otherwise catastrophic resonance amplitudes in large scale structures as skyscrapers [41, p.210]. On the microscale, few reports on mechanical systems with these characteristics are available. A corresponding asymmetric system with identical resonance frequencies as a method to amplify the amplitude of a smaller resonator by driving it with a larger one with the same resonance frequency was discussed by Li et al. [216, 217]. They employed this method to induce higher amplitudes in a small microscaled oscillator system but also detected the resulting amplitude response at this small system. There are also other experiments that employ resonantly coupled nanoscaled mechanical resonators to different oscillating systems as for example optical cavity fields [218] or an electronic spin system [219].

In this chapter a theory is developed to understand and employ an easy-to-detect oscillating micromechanical system that is mechanically coupled to a very small, otherwise difficult to detect but sensitive nanomechanical resonator to remotely sense the influence of a force gradient on this nanomechanical resonator.

A single report by Torres et al. is available in a conference proceedings in which they observe an enhanced frequency shift signal of an asymmetric coupled cantilever system in response to mass changes [220]. In this experiment the cantilever systems oscillatory state is detected capacitively on the larger cantilever [220]. The cantilevers in this report still have very different resonance frequencies and the smaller cantilever is large enough that both modes of the coupled system are clearly observable. While Torres et al. already discussed the potential of such an approach to create more sensitive mass sensors, they suggest neither a corresponding theory nor any possibilities for improvements aside from the use of smaller oscillators. As will be demonstrated below simply decreasing the mass or spring constant of the smaller oscillator is not sufficient to obtain sensors with an enhanced frequency shift signal. It is vital that the resonance frequencies of the subsystems are close to each other in order to obtain a sensor with a high frequency shift signal in response to an external force gradient or mass change.

6.2 Coupled harmonic oscillator model

It was discussed in chapter 2, that the driven harmonic oscillator is a very good description of the measurement principle of dSFM sensors. Therefore, this discussion of coupled oscillator sensor systems is also based on a comparatively simple coupled harmonic oscillator model as shown in figure 6.1. This model contains two oscillator subsystems as well as an additional spring and damping element to model the interaction with the environment. Furthermore, the system is driven by a harmonic force $F(t) = F_0 \cos(\omega_D t)$ applied to the first subsystem.

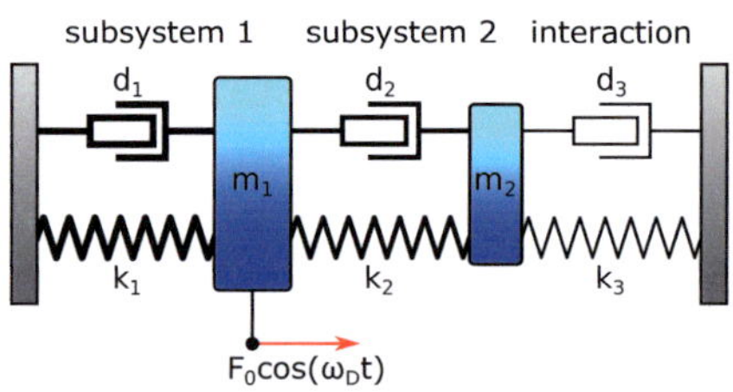

Figure 6.1: Driven coupled harmonic oscillator model with two subsystems and external interaction

6.2.1 Equations of motion and analytical solutions

The equation of motion of such a coupled system can be written in matrix form. Each subsystem consists of a mass m_i, a spring with spring constant k_i and a damping element with damping factor d_i, with $i \in \{1, 2\}$ being the number of the subsystem. An external force gradient and dissipation are described by the constants k_3 and d_3 respectively. With the deflection vector $\vec{s}(t) = (s_1(t), s_2(t))$ with s_i being the deflection of mass m_i from its equilibrium position, the system of differential equations to describe the oscillator system's motion is given by the expression

$$\mathbf{M} \cdot \ddot{\vec{s}}(t) + \mathbf{D} \cdot \dot{\vec{s}}(t) + \mathbf{K} \cdot \vec{s}(t) = \vec{F}(t) \tag{6.3}$$

with the matrices

$$\mathbf{M} = \begin{pmatrix} m_1 & 0 \\ 0 & m_2 \end{pmatrix}, \quad \mathbf{D} = \begin{pmatrix} d_1 + d_2 & -d_2 \\ -d_2 & d_2 + d_3 \end{pmatrix}, \quad \mathbf{K} = \begin{pmatrix} k_1 + k_2 & -k_2 \\ -k_2 & k_2 + k_3 \end{pmatrix} \tag{6.4}$$

and the excitation force vector $\vec{F}(t) = (F_0, 0) \cos(\omega_D t)$.

This kind of model is often discussed in basic textbooks on classical mechanics [27, 28]. However, solutions to the above system of differential equations are commonly derived for simplified undamped or symmetric systems. In order to describe the behavior of a strongly asymmetric system and its response to external force gradients, however, the complete solutions are necessary. Similar to the simple harmonic oscillator the full solutions to the differential equations consist of two parts. The homogenous part is important when describing transient states of the system but will diminish over time due to damping. The particular part of the solution corresponds to the steady state, which is typically the one of interest for dSFM and other sensor applications.

This particular solution was derived by employing a complex ansatz $\vec{\hat{s}}(t) = \vec{\hat{A}} \cdot \exp(i\omega_D t)$ with a complex time-independent amplitude vector $\vec{\hat{A}} = (\hat{A}_1, \hat{A}_2)$ and by rewriting the excitation force as a complex vector $\vec{\hat{F}}(t) = (F_0, 0) \cdot \exp(i\omega_D t)$. The time derivatives of this ansatz were calculated and used in equation 6.3. This resulted in a system of linear equations for the complex amplitudes. The resulting coefficient matrix was solved by elementary row operations to obtain the complex amplitudes of the subsystems. These were separated in real amplitudes and phases $(A_1, A_2, \phi_1, \phi_2)$ that are each functions of the various parameters of the oscillator system including the driving frequency and the external spring and damping elements. The stationary solution to the system of differential equations 6.3 is then given by the real part of the complex ansatz $\vec{s}(t) = (A_1 \cos(\omega_D t + \phi_1), (A_2 \cos(\omega_D t + \phi_1))$. While these analytical solutions were used for the calculations in the remainder of this chapter, their exact expressions are quite cumbersome and are therefore not shown here. They can be found in appendix A instead.

Parameter	Subsystem 1	Subsystem 2
Spring constants	$k_1 = 1\ N/m$	$k_2 = 0.003$ N/m
Quality factors	$Q_1 = 5000$	$Q_2 = 500$
Eigenfrequency	$\omega_1 = 500000$ 1/s	$\omega_2 =$ varying
Driving amplitude	$F_0 = 10^{-10}$ N	-

Table 6.1: Parameter set used for the calculations involving the coupled harmonic oscillator model in this chapter

The derived solutions are valid for any combination of parameters. The aim of this chapter is the proposal of a new sensor concept based on the combination of a microstructured cantilever and a nano-sized resonator. Therefore, the following discussion is restricted to an asymmetric parameter set with subsystem 1 representing a bigger oscillator and subsystem 2 a smaller one with $k_1 >> k_2$ and $m_1 >> m_2$. Furthermore, as the main topic of this work was the development of novel dSFM sensors, the treatment of the system's interaction in the following is limited to an external force gradient, e.g. $d_3 = 0$ and the masses are assumed to be constant. Additionally it is favorable to retain the easy determination of the oscillatory state of the bigger oscillator. Hence, the observables of the bigger system are in the focus of interest.

The exemplary parameter set used for the calculations within this chapter is summarized in table 6.1. These parameters were chosen to resemble those of a cantilever-FeCNT system. The masses $m_i = \omega_i^2/k_i$ of the partial oscillators were always calculated according to the spring constants and the eigenfrequencies. This resulted in constant spring constants of the subsystems throughout the calculations. The damping factors $d_i = k_i/(\omega_i Q_i)$ of the systems were derived from the experimentally more accessible quality factors Q_i (see equation 2.9).

6.2.2 General behavior

The behavior of a coupled harmonic oscillator system with two masses can be described by the formation of two oscillation modes with respective resonance frequencies. In case of a strongly asymmetric system with very dissimilar eigenfrequencies, however, the influence of the smaller system on the movement of the bigger system is usually very small and the second resonance mode will be extremely difficult to detect in the oscillation of the bigger system. For example, a calculation with the exemplary parameters defined above and an eigenfrequency of the smaller subsystem of $\omega_2 = 2\omega_1$ revealed that the amplitude of the second mode is only visible in the amplitude versus driving frequency curve of the big oscillator as a tiny feature with an amplitude change that is more than four orders of magnitudes smaller than the resonance amplitude of the first mode. Furthermore, the resonance frequencies of the two modes were also found to be approximately equal to those of the two individual subsystems.

This changed drastically, however, if the eigenfrequencies and, connected to this for the case of low damping, the resonance frequencies of the two subsystem were close to each other. Then both resonance modes were clearly visible in the amplitude and phase versus driving frequency curves of the big oscillator as shown in figure 6.2. Additionally, the resonance frequencies of the two modes were found to be different from those of the subsystems. The jump in the phase back to almost zero between the two resonance peaks is connected to a strong dip of the amplitude at a given frequency. This is a typical feature of driven coupled systems sometimes refereed to as 'anti-resonance' [221]. Furthermore, the two resonance peaks of the coupled system never cross each other and maintain a certain minimal frequency separation that depends on the spring constant of the smaller subsystem, which couples the two masses [215, 222].

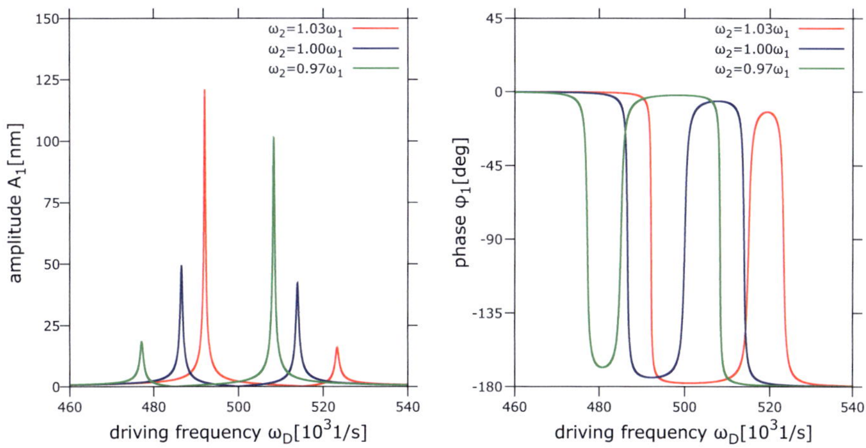

Figure 6.2: Amplitude and phase versus driving frequency curves of the bigger subsystem of the coupled harmonic oscillator model for various eigenfrequency ratios of the subsystems.

6.2.3 Frequency shifts in response to external force gradients

With the aim to propose a novel dSFM force sensor concept, the changes of the oscillatory state of the system with respect to an external force gradient k_3 acting on the smaller oscillator have to be studied. In standard dSFM, either a shift of the eigenfrequencies of the modes or the corresponding change of the amplitude of the oscillation at a given excitation frequency can be used as a measurement signal. Since the dSFM experiments in this work were conducted in high vacuum with the frequency modulation technique the following discussion focuses on the frequency shifts of the resonance modes induced by external force gradients. Nevertheless, the change in the amplitude of the sensor induced by an external force gradient k_3 is also briefly reviewed.

Eigenfrequencies of the coupled system

In order to derive a frequency shift signal for the coupled system for a given set of parameters either the shift of the eigenfrequencies of the system or the shift of the resonance frequencies can be investigated. In close analogy to the derivation of the frequency shift of the simple harmonic oscillator in section 2.2.3, the case without damping is considered. The eigenfrequencies of the system can be written as [27, p.213]:

$$\omega_{a/b,k_3}^2 = \frac{\omega_{12}^2 + \omega_{23}^2}{2} \mp \sqrt{\frac{\left(\omega_{12}^2 - \omega_{23}^2\right)^2}{4} + \frac{k_2^2}{m_1 m_2}} \tag{6.5}$$

with $\omega_{ij} = \sqrt{(k_i + k_j)/m_i}$ being the combined eigenfrequencies of the subsystems, i.e. the eigenfrequencies the masses of the subsystem would exhibit if the other mass is kept stationary. This equation shows a minimal frequency seperation of the eigenfrequencies of the coupled system with a value of $2\sqrt{k_2^2/(m_1 m_2)}$ in the case $\omega_{12}^2 = \omega_{23}^2$. According to equation 6.5 the index a denotes the mode with the lower eigenfrequency and b the mode with the higher eigenfrequency accordingly in the following.

With equation 6.5 the shift of the eigenfrequencies of the modes of the coupled oscillator for a given external force gradient $k_3 = \Delta k$ can be calculated according to:

$$\Delta\omega_{a/b} = \omega_{a/b,\Delta k} - \omega_{a/b,0}. \tag{6.6}$$

The magnitudes of these frequency shifts do not only depend on the external force gradient and the spring constants of the subsystems but also on the relative values of the eigenfrequencies of the subsystems. This is visualized in figure 6.3 that shows the frequency shifts of the system for a fixed value of the external force gradient $\Delta k = 10^{-6}$ N/m depending on the eigenfrequency ω_2 of the smaller subsystem.

Not considering the case of very close frequency matching, the mode with the eigenfrequency closer to ω_2 was found to exhibit a larger frequency shift signal in the calculations, so it would be desirable to use this mode for measurement purposes. This does not consider the actual amplitudes of the modes, however, and usually the mode with the larger

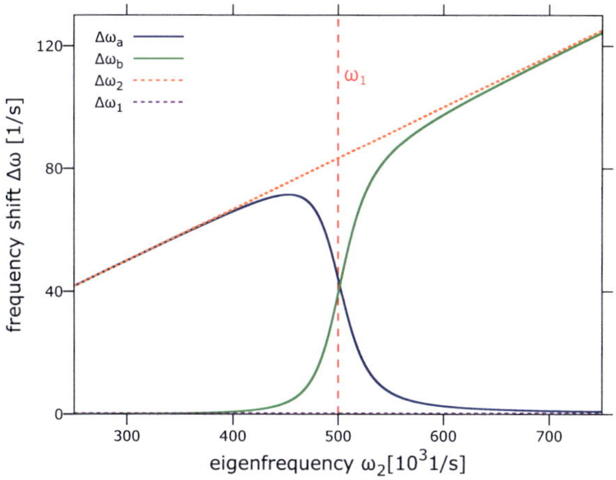

Figure 6.3: Frequency shift signal of both oscillation modes of the coupled system depending on the resonance frequency of the smaller subsystem. The external force gradient was given by $\Delta k = 10^{-6}$ N/m. The frequency shift signals of the subsystems $\Delta\omega_i = \omega_i/(2k_i)\Delta k$ when treated as individual harmonic oscillator sensors are indicated.

frequency shift was found to be the one with the lower maximal amplitude when observed on the bigger subsystem. A lower amplitude means that the corresponding mode is more difficult to detect and usually the mode with the higher maximal amplitude is used for an actual measurement. Nevertheless, if the eigenfrequencies of the system were close to each other, the frequency shift in response to an external force gradient of this higher amplitude mode was found to be much larger than expected for the bigger oscillator alone. This can be regarded as a transfer of part of the small subsystem's high sensitivity to the bigger oscillator by the mechanic coupling. If the eigenfrequency of the smaller subsystem was set to be much higher than that of the bigger subsystem, the frequency shift signals of the modes were found to be approximately given by those of the individual subsystems. The corresponding maximal amplitude of the more sensitive mode with higher eigenfrequency is then usually negligibly small in the amplitude versus driving frequency curves of the bigger subsystem. For a real system, this means that only one resonance peak is visible in the amplitude versus driving frequency curve and this corresponds to the case of a rigid smaller subsystem as was discussed in chapter 5 for the FeCNT interaction tips of the bidirectional sensors.

The eigenfrequencies of the undamped coupled system can be easily calculated. Unfortunately, they are not accessible experimentally since they do not correspond to the frequencies at which the cantilever's phase with respect to the excitation is given by $\phi_1 = -\pi/2$ as in the case for a simple harmonic oscillator. The experimentally accessible frequencies are the resonance frequencies, i.e. the frequencies at which the amplitude response of the system shows a maximum and the frequencies at which the phase with respect to the excitation is given by a fixed value $\phi_1 \in (0, -\pi)$. If these resonance frequencies or the frequencies corresponding to specific phases of the big oscillator and their frequency shifts in response to external force gradients are required, they can be derived by using the analytical formulas for the amplitudes $A_i(\omega_D)$ and phases $\phi_i(\omega_D)$ in appendix A. Usually the simplest way to determine these frequency is given by employing numerical methods to obtain high precision results. In the limit of vanishing damping, however, these frequencies coincide with the eigenfrequencies of the coupled system as is the case for the simple harmonic oscillator [223, p.41]. This means that in the case of low damping, on the one hand the eigenfrequencies are a good approximation for the experimentally accessible distinguished frequencies and on the other hand a matching of the eigenfrequencies can be approximated by a matching of the experimentally accessible resonance frequencies.

For verification a numerical study using the software 'Mathematica 10.0' was conducted using the analytical formulas for the amplitudes and phases. The calulation was based on the coupled system with low damping, as given by the parameters summarized in table 6.1 and wide ranges of the parameters ω_2 and k_3 . The resonance frequencies and the frequencies at which $\phi_1 = -\pi/2$ holds and their shifts in response to the external force gradient k_3 were calculated. The relative deviation of these frequencies from the eigenfrequencies and the relative deviation of the corresponding frequency shifts of the undamped system were found to be in the order of 0.1%. This validates the good description of the frequency shifts of the coupled system by the eigenfrequencies.

Amplification factors

Although the calculation of the frequency shift signal for any external force gradient is possible with equation 6.6 as shown in figure 6.4, usually a more simple expression is favorable. Moreover, the frequency shift signal is the observable and the external force gradient is the quantity of interest. Therefore an equation allowing for an easy inversion is more desirable for practical applications.

As can be seen in figure 6.4 the dependance of the frequency shift for small absolute values of the external force gradient $|\Delta k|$ was found to be approximately linear. Therefore,

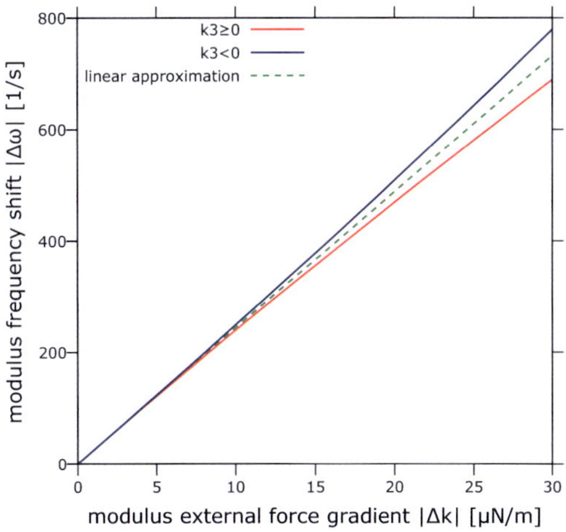

Figure 6.4: Modulus of the frequency shift $\Delta\omega$ of the coupled system depending on the modulus of the external force gradient Δk. A fixed value of $\omega_2 = 1.03\omega_1$ for the eigenfrequency of the smaller subsystem was employed. The frequency shift was calculated for the peak with the higher amplitude as shown in figure 6.2. The result of the linear approximation given by equation 6.7 with the calculated amplification factor according to equation 6.8 is indicated.

an approximation of the frequency shift for small $|\Delta k| << k_2$ can derived in analogy to equation 2.11 for a single harmonic oscillator:

$$\Delta\omega_{a/b} \approx \frac{\omega_{a/b,0}}{2k_{a/b,\Lambda}}\Delta k = \frac{\omega_{a/b,0}\Lambda_{a/b}}{2k_1}\Delta k \tag{6.7}$$

with the enhanced effective spring constant $k_{a/b,\Lambda} = k_1/\Lambda_{a/b}$ for each mode. These effective spring constants can be described in terms of amplification factors $\Lambda_{a/b}$ for each of the two modes respectively that indicate the enhancement of the signal of the big oscillator by the coupling to the smaller more sensitive oscillator system as compared to the single harmonic oscillator case.

The expressions for the amplification factors were derived by a linearization of equation 6.5 by employing Taylor expansions and omitting terms of higher orders of k_3/k_2:

$$\Lambda_{a/b} \approx \frac{k_1}{k_2}\frac{\alpha^2 \mp \frac{\alpha^4-\alpha^2}{\zeta}}{1+\alpha^2 \mp \zeta}. \tag{6.8}$$

This equation contains the ratio α of the eigenfrequency of the smaller system to the first combined eigenfreqency $\omega_2 = \alpha\omega_{12} = \alpha\omega_1\sqrt{(1 + (k_2/k_1))}$ and the definition

$$\zeta^2 := 1 + 2\alpha^2 \left(\frac{2k_2}{k_1 + k_2} - 1 \right) + \alpha^4. \tag{6.9}$$

These amplification factors $\Lambda_{a/b}$ only depend on the spring constants and resonance frequencies of the subsystems. The approximations leading to equation 6.8 are valid for small external force gradients $k_3 = \Delta k$ as long as the expressions

$$|k_2| >> |k_3| \tag{6.10}$$

$$\left| \frac{k_2\zeta^2}{2\alpha^4 - 2\alpha^2} \right| >> |k_3| \tag{6.11}$$

$$\left| \frac{k_2\sqrt{1 + \alpha^2 \mp \zeta}}{\alpha^2 \mp \frac{2\alpha^4 - 2\alpha^2}{2\zeta}} \right| >> |k_3| \tag{6.12}$$

are satisfied. The complete derivation of these formulas can be found in appendix B.

From a practical point of view an interesting quantity is the sensor amplification factor Λ that corresponds to the amplification factor of the mode with the larger amplitude without an external force gradient. Please note, that for a system with a very small initial eigenfrequency separation, an external force gradient may result in a change of the role of the 'mode with the larger amplitude'. The sensor amplification factor was found to increase drastically for diminishing frequency separation of the eigenfrequencies as is shown in figure 6.5. This also demonstrates that a signal increase by more than two orders of magnitude is easily attainable.

6.2.4 Amplitudes of the coupled system

The discussion of the frequency shifts contained no information of the corresponding changes of the amplitudes of the involved modes. With the analytical formulas for the amplitudes $A_i(\omega_D)$ it is possible to calculate the response of the big oscillator to external force gradients acting on the small oscillator. Such a calculation is shown in figure 6.6 and demonstrates the effect of the external force gradient as the resonance frequency and amplitude changes of both resonance peaks, which can be used as measurement signals, can be clearly seen. Moreover, the already mentioned larger frequency shift as well as the far smaller amplitude of the second peak is visible.

If the resonance frequencies of the subsystems are close together the smaller oscillator is essentially driven close to its resonance frequency by the already resonantly amplified

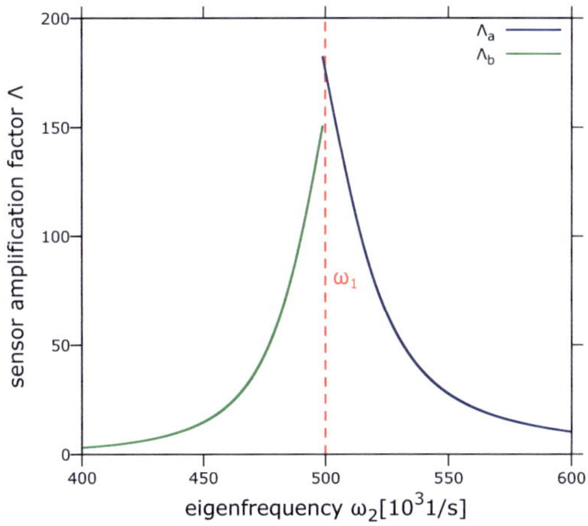

Figure 6.5: Sensor amplification factor Λ depending on the eigenfrequency of the smaller subsystem ω_2. The two branches correspond to the amplification factor of the mode with the higher amplitude. An sufficiently small external force gradient of $\Delta k = 10^{-6}$ N/m was used for the calculation. Please note that the eigenfrequency of the smaller subsystem at which the resonance amplitude of the two modes are of equal height does not coincide with the point of intersection of the amplification factor curves of the two modes resulting in a discontinuity.

amplitude of the bigger oscillator. Hence, its amplitude is larger than in the case of far separated resonance frequencies. The actual amplitude amplification factor can be calculated if the spring constants, resonance frequencies and damping coefficients of the subsystems are known. This amplification of the resonance amplitude of the smaller system is depicted in figure 6.7 for the exemplary system given by table 6.1 and corresponds to the already mentioned observations of Li et al. [216, 217].

6.3 Application to force sensors

The discussed asymmetric coupled harmonic oscillator model showed a surprisingly complex behavior. The results from this discussion can be employed to create a new class of micro- and nanomechanical sensor designs. By mechanically coupling two strongly asymmetric mechanical oscillator systems and manipulating their resonance frequencies to be close to each other, the resulting combined sensor system retains the ease-of-detection of the oscillatory state of a comparatively big oscillating subsystem with a large fraction of the high force gradient sensitivity of a nanoscaled resonator. For prospective sensors based on

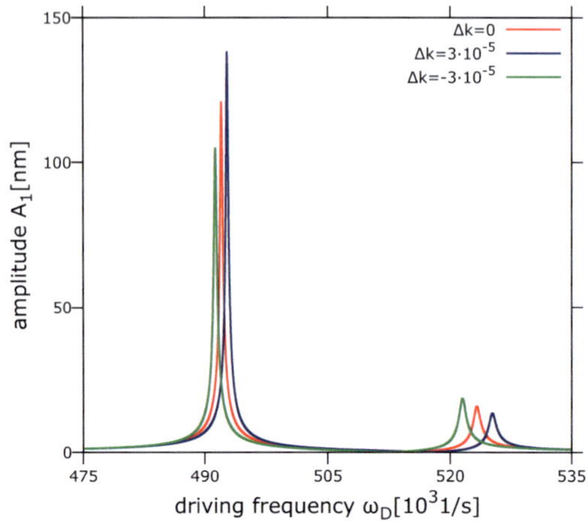

Figure 6.6: Amplitude response of the bigger subsystem of the coupled system for different external force gradients. The eigenfrequency of the smaller subsystem that was used for the calculation is given by $\omega_2 = 1.03\omega_1$.

this concept a matching of the resonance frequencies can be implemented by either changing the effective masses or spring constants of both subsystems. To stress the importance of closely matched resonance frequencies, which leads to a notable combined resonant oscillation of the subsystems, this state will be called 'co-resonant' in the following.

As the underlying harmonic oscillator model is of a very general nature it can be applied to many different sensors based on resonant mechanical oscillation. It is emphasized that although the above considerations focused on the application as force gradient sensors a similar kind of discussion can be conducted to include mass changes. This can then be used to describe co-resonant mass-sensors.

The signal increase of the sensor is accompanied by the mentioned amplification of the amplitude of the smaller subsystem and it has to be kept in mind that in a real system high amplitudes can lead to non-linear behavior of the components. Furthermore, it has also to be mentioned that the steady state resonance amplitude relationship between the two oscillator subsystems was found to depend on the external spring constant k_3 by using the analytical formulas for the amplitudes A_i. Therefore, the amplitude of the smaller subsystem still changes in response to a change of this external spring constant even if the amplitude of the bigger subsystem is kept constant by the measurement system. To what extend this leads to a transient behavior of the coupled system and if this influences the

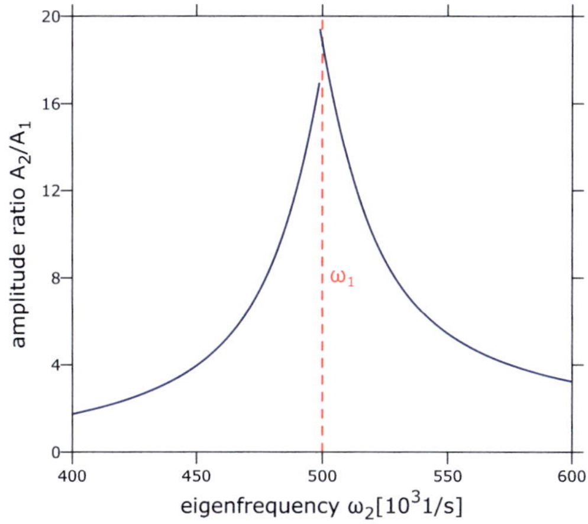

Figure 6.7: Amplitude ratio between the amplitudes of the bigger and the smaller oscillator depending on the eigenfrequency of the smaller subsystem. The ratio was calculated for the mode with the larger amplitude of the bigger oscillator at the corresponding resonance frequency.

possible measurement speed are questions of practical relevance and should to examined in future work.

Nevertheless, this sensor concept provides the means to overcome the low signal of the lateral mode of bidirectional dSFM sensors as will be discussed in the next chapter.

Very recently, after the corresponding patent application was filed [224] and the experiments presented in this work were concluded, another report on a coupled sensor system was published by Vidal-Álvarez et al. [225]. In addition to an experimental implementation with a signal increase by a factor of approximately two they presented elaborate numerical calculations on a continuous coupled cantilever-nanowire model for mass sensing. These calculations focused on the nanowire length as the free parameter and while they also demonstrated a corresponding signal enhancement, the underlying influence of the resonance frequencies of the subsystems on the signal amplification was not discussed. Furthermore, they discuss only one of the two oscillation modes of the system. Nevertheless, this publication is an additional indicator that the more general sensor concept proposed in this chapter is viable.

7 Co-resonant signal enhancement for scanning force microscopy sensors

With the co-resonant sensor concept described in the previous chapter it is possible to create highly sensitive dSFM sensors. The bidirectional sensors discussed in chapter 5 already contain the main components required for a co-resonant signal enhancement: a microscaled cantilever acting as the big oscillator and the iron-filled carbon nanotube that can be employed as a nanoscaled resonator with high force gradient sensitivity [6].

7.1 Implementation of the co-resonant sensor concept

As already discussed in section 4.4.1 carbon nanotubes are employed as tips for SFM sensors due to their advantageous physical properties [125]. Typically, they are regarded as a static part of the sensor, however.

Publications examining the influence of the CNT tip's properties on the dynamic behavior of the SFM sensor are usually only concerned with the quasi-static elastic response of the CNT during a tapping mode measurement and the influence of a possible bending or buckling of the CNT on the SFM sensor's behavior [127, 128, 226].

To the author's knowledge, no dSFM experiments that employ the dynamic properties of a sensitive carbon nanotube resonator have been published to date.

The bidirectional dSFM sensors with the FeCNT interaction tips can be considered a coupled harmonic oscillator system with dissimilar resonance frequencies of the partial oscillators. According to the co-resonant sensor concept, matching the resonance frequencies of the second flexural mode of the cantilever and the first flexural mode of the carbon nanotube should result in a co-resonant bidirectional sensor with an enhanced signal in the lateral oscillation mode.

Please note that the spacer element of a bidirectional sensor could be regarded as a third possible oscillator. Since its resonance frequency is usually far above that of both cantilever and iron-filled carbon nanotube, however, it is simply considered as a stiff connection between the two oscillator systems in the following.

To demonstrate the applicability of the co-resonant sensor concept to the bidirectional sensors, a special bidirectional sensor was fabricated: To ensure an already reasonably low resonance frequency of the nanotube resonator, a very long FeCNT was chosen and attached to the sensor's spacer element.

7.1.1 Sensor characterization

For the description of the properties of a bidirectional sensor with co-resonant sensitivity enhancement, not only the dynamic properties of the cantilever but also that of the FeCNT have to be considered.

These properties of the FeCNT were determined by vibration measurements inside the SEM equipment by employing the vibration stage after attachment of the FeCNT to the bidirectional sensor's spacer element. The resonance frequency $f_{1,FeCNT}^{*,p}$ and the Q factor of the FeCNT were determined by measuring the amplitude of the free end of the FeCNT depending on the frequency of the harmonic excitation. The parameters were then extracted from the data by fitting the amplitude versus driving frequency data with the amplitude expression containing the Q factor from equation 2.10 with the fit parameters $A_0, Q = Q_{1,FeCNT}, \omega_0 \approx 2\pi f_{1,FeCNT}^{*,p}$.

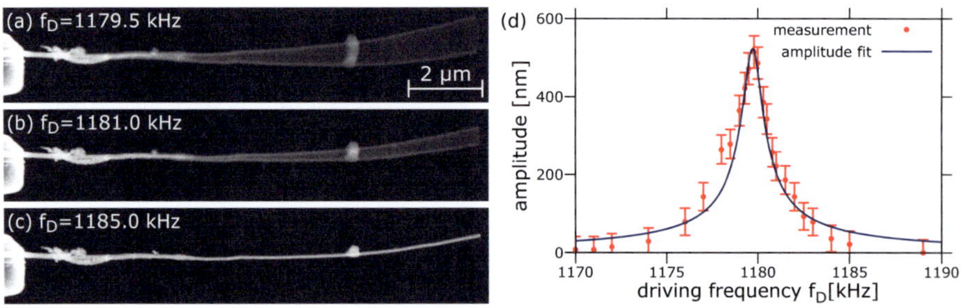

Figure 7.1: Measurement of the resonance frequency and quality factor of an iron-filled carbon nanotube used as interaction tip for a co-resonant sensor: (a)-(c) SEM images at various excitation frequencies f_D, (d) measured amplitudes depending on the excitation frequency and corresponding amplitude curve based on the fitted parameters $f_{1,FeCNT}^{*,p} = (1179.71 \pm 0.25)$ kHz, $Q_{1,FeCNT} = (1.07 \pm 0.38) \cdot 10^3$ and $A_0 = (0.49 \pm 0.11)$ nm.

Corresponding exemplary SEM images from the measurement process are depicted in figure 7.1(a)-(c). The resulting amplitude versus driving frequency curve can be found in figure 7.1(d). In the previous experiments presented in this work, the length measurements in the SEM were always spanning many pixels of the image, therefore the error induced by the limited pixel resolution of the images was negligible. In the present case, the measured amplitudes often corresponded to only a few pixels, hence this was considered in the measurement errors. The corresponding fit with the software Origin 8.6 resulted in the parameter set $f^{*,p}_{1,FeCNT} = (1179.71 \pm 0.25)$ kHz, $Q_{1,FeCNT} = (1.07 \pm 0.38) \cdot 10^3$ and $A_0 = (0.49 \pm 0.11)$ nm. For the fit, squared amplitudes were used to circumvent numerical difficulties when fitting the amplitude formula, which contains a square root. The given margins of error of the parameters correspond to the $\pm 3\sigma$ environment, based on the standard errors calculated by the fit software.

The FeCNT can also be considered a beam like structure. Therefore, the dynamic spring constant of its first flexural mode can be calculated the same way as for the cantilever with equations 2.33 and 2.37. The required density of the FeCNT

$$\rho_{FeCNT} = \rho_{Fe}\frac{d^2_{Fe}}{d^2_{FeCNT}} + \rho_C\frac{d^2_{FeCNT} - d^2_{Fe}}{d^2_{FeCNT}} \tag{7.1}$$

can be approximated as a mean density taking into account the volume ratios of the materials in the FeCNT. Here, d_{FeCNT} is the diameter of the whole FeCNT and d_{Fe} that of the iron filling. For the density of carbon the literature value of highly ordered pyrolitic graphite $\rho_C = 2266$ kg/m^3 [95, p.22][227] can be used to emulate the carbon shells. Please note that this derivation of the spring constant of the iron-filled carbon nanotube does not regard the attachment of the FeCNT as a torsional spring as suggested in previous works [6, 115]. However, compared to the experiments described in these publications, the attachment procedure has been improved by using a gas injection system to supply a carbon precursor instead of relying on residual hydrocarbons from inside the SEM system. This resulted in a stiff behavior of the deposited carbon attachments as can be seen in figure 7.1(b). The approximation for the density in equation 7.1 furthermore neglects the carbon caps at the end of the FeCNT and deviations from a cylindrical shape.

Examples for such deviations are visible in figure 7.1(c). Here, two attached particles are visible, one near the point of attachment and one about halfway down the FeCNT. Also, a thicker carbon layer is visible near the point of attachment. The thicker carbon layer most likely consists of amorphous carbon as has been observed on similar FeCNTs [93]. Amorphous carbon has a much lower elastic modulus than the carbon nanotubes [93], therefore, the influence of this carbon layer on the effective bending stiffness of the FeCNT was neglected. Furthermore, the influence of added masses that are located close to the attachment of a cantilever beam-like structure on the effective mass and, consequently the resonance frequency of the the whole system is rather small, as they are not accelerated to high speeds during an oscillation [228]. Hence, the masses of the thicker carbon layer and the corresponding particle close to the attachment were also neglected when calculating the spring constant using the resonance frequency according to equation 2.37.

In contrast, a particle further along the length of a cantilever beam-like structure is accelerated to comparatively high speeds when the system is excited at is fundamental resonance frequency. This strongly influences the resonance frequency of the system [228], and therefore, the particle at the second half of the FeCNT visible in figure 7.1(c) had to be taken into account for the calculation of the spring constant of the iron-filled carbon nanotube. There is no notable material contrast between this particle and the shell of the FeCNT. As the FeCNT sample should only contain carbon, iron and corresponding compounds [93, 94], the particle was assumed to consist of carbon. A particle mass of $m_p = (31.4 \pm 2.6)$ fg was deduced by considering a plate-like geometry and measuring the thickness and the lateral cross sectional area of the particle in the SEM at various angles. The mass of the FeCNT was calculated by assuming a cylindrical shape according to the following equation:

$$m_{FeCNT} = L_{FeCNT} \left(\rho_{Fe} \frac{\pi d_{Fe}^2}{4} + \rho_C \frac{\pi \left(d_{FeCNT}^2 - d_{Fe}^2 \right)}{4} \right). \tag{7.2}$$

Here, L_{FeCNT} is the length of the FeCNT. The geometric parameters were obtained from SEM images and a value of $m_{FeCNT} = (294 \pm 63)$ fg was found. Hence, the mass of the particle is considerably lower than the mass of the FeCNT. Additionally the spatial extent of the particle is small. Therefore, its influence on the oscillation of the FeCNT was considered as that of an added point mass with an negligible effect on the mode shape of the FeCNT. In such a case the resonance frequency of the FeCNT $f_{1,FeCNT}^*$ without this particle can be approximated by [228]:

$$f_{1,FeCNT}^* = f_{1,FeCNT}^{*,p} \sqrt{1 + \frac{m_p}{m_{FeCNT}} (u_{1,FeCNT}^N(x_p))^2} \tag{7.3}$$

with $u_{1,FeCNT}^N(x)$ being the mode shape of the iron-filled carbon nanotube according to the Euler-Bernoulli beam theory, normalized to satisfy

$$\int_0^{L_{FeCNT}} (u_{1,FeCNT}^N)^2 dx = L_{FeCNT}. \tag{7.4}$$

The position of the center of the particle along the length of the FeCNT $x_p = (8.90 \pm 0.27)$ µm was measured with the SEM and the required normalization of the mode shape was calculated numerically. With these values the resonance frequency of the carbon nanotube without the particle $f_{1,FeCNT}^* = (1340 \pm 73)$ kHz was obtained according to equation 7.3. With this corrected resonance frequency, the mean density from equation 7.1 and the measured geometrical parameters, finally the spring constant of the FeCNT could be calculated with equations 2.33 and 2.37. The result of this calculation along with the obtained relevant properties of the bidirectional sensor and it's FeCNT as well as their corresponding margins of error can be found in tables 7.1 and 7.2.

The Q factor of the cantilever was obtained according to equation 2.9 by using the resonance frequency of the cantilever and measuring the full bandwidth of the corresponding resonance peak in an amplitude response curve determined with the dSFM equipment. Even though there are more precise ways to determine the Q factor of a cantilever sensor, as for example the 'ring-down' method [8, pp.81-86], they require more complex measurements.

Length	$L_{FeCNT} = (12.01 \pm 0.36)$ µm
Diameter	$d_{FeCNT} = (111 \pm 7)$ nm
Iron filling diameter	$d_{Fe} = (24 \pm 5)$ nm
Resonance frequency first flexural mode	$f^*_{1,FeCNT} = (1340 \pm 73)$ kHz
Q factor first flexural mode	$Q_{1,FeCNT} = (1.07 \pm 0.38) \cdot 10^3$
Dynamic spring constant	$k^{dyn}_{FeCNT,1} = (5.2 \pm 1.7)$ mN/m

Table 7.1: Relevant geometrical and oscillation properties of the iron-filled carbon nanotube of the bidirectional sensor used for the experiments to demonstrate the co-resonant enhancement. Please note that the resonance frequency indicates the calculated value without the added particle.

Length	$L = (225.8 \pm 6.8)$ µm
Cross sectional area	$S = (67.8 \pm 4.1)$ µm^2
Effective spacer length	$l_{dist} = (23.1 \pm 0.7)$ µm
Resonance frequency first flex. mode	$f^*_1 = (66.90 \pm 0.01)$ kHz
Resonance frequency second flex. mode	$f^*_2 = (422.13 \pm 0.01)$ kHz
Q factor second flex. mode	$Q_2 = (3.46 \pm 0.11) \cdot 10^3$
Perpendicular dynamic spring constant	$k^{dyn}_{x_{tip},1} = (3.19 \pm 0.30)$ N/m
Lateral dynamic spring constant	$k^{dyn}_{\Delta x,2} = (337 \pm 72)$ N/m

Table 7.2: Relevant geometrical and oscillation properties of the cantilever of the bidirectional sensor used for the experiments to demonstrate the co-resonant enhancement.

7.1.2 Resonance frequency adjustment

According to the co-resonant sensor concept, the frequency shift signal of the big oscillator improves if the resonance frequencies of the two partial systems are close to each other.

In the case of the bidirectional sensors used in this work, the resonance frequency of the iron-filled carbon nanotubes was usually found to be far above that of the second flexural mode of the cantilever. It is difficult to modify the corresponding resonance frequency of a cantilever without changing the position of the nodal point, which would impair the functionality of a bidirectional sensor. Therefore, it is more feasible to modify the resonance frequency of the FeCNT.

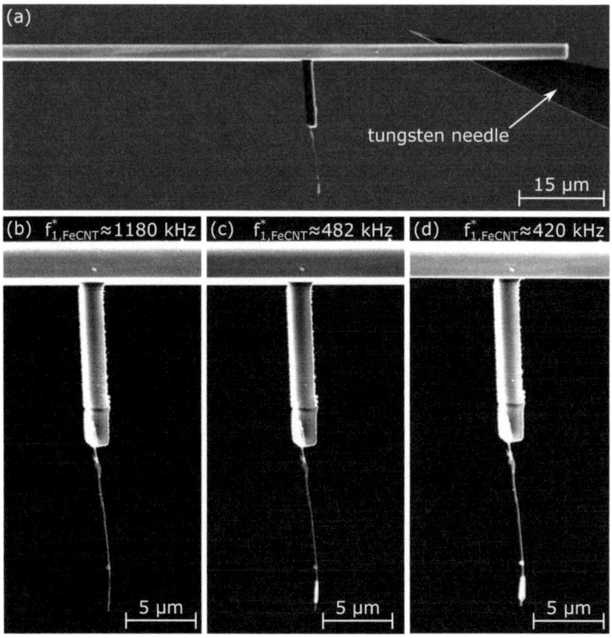

Figure 7.2: Resonance frequency adjustment of a iron-filled carbon nanotube by electron beam induced material deposition: (a) the cantilever's free end was pinned by a tungsten needle to allow for an observation of the resonance frequency of the iron-filled carbon nanotube, (b)-(d) the iron-filled carbon nanotube with the deposited material at various stages of resonance frequency adjustment. Please note that the image in (d) was taken at a slightly different viewing angle compared to (b) and (c).

For the bidirectional sensor sensor used in these verification experiments, electron beam induced material deposition was used to create an additional mass near the free end of the FeCNT, thus increasing it's effective mass and reducing its resonance frequency. During the mass deposition the resonance frequency of the FeCNT was monitored by using the SEM and the vibration stage. To ensure that the observed resonance frequency is actually that of the FeCNT alone and not a combined mode of the FeCNT-cantilever system, the free end of cantilever was pinned by using the micromanipulator with a tungsten needle as shown in figure 7.2 (a). This detuned the cantilever's resonance frequency and thereby induced a stiff behavior of the cantilever at the resonance frequency of the FeCNT, which made it possible to adjust the resonance frequency of the FeCNT to a desired value. The adjustment was conducted in several steps: at the beginning a platinum precursor was used as the higher density of the deposited material reduced the resonance frequency of the FeCNT quickly. When the resonance frequency of the FeCNT was measured to be roughly 120% of the cantilever's second flexural mode, the gas source was switched to the carbon precursor, since the resulting material deposition changed the resonance frequency of the FeCNT more slowly. SEM images of the FeCNT at different stages of the deposition process can be found in figures 7.2(b) to (d). Before the deposition, the FeCNT was behaving rigidly when the system was excited close to the resonance frequency of the second flexural mode of the cantilever as can be seen in figure 7.3(a). When the resonance frequency of the FeCNT approached that of the cantilever both systems were observed to exhibit collective resonant oscillations as expected. This is shown in figure 7.3(b). For very close frequency matching very small excitation amplitudes had to be used in order to not damage the FeCNT due to the amplitude amplification. In these cases the oscillation of the cantilever was no longer visible as shown in the SEM image in figure 7.3(c). Of course, to obtain these images the cantilever was not pinned by the tungsten needle. These observations already demonstrate the co-resonant amplitude amplification.

7.2 Experimental demonstration of the co-resonant signal enhancement

To gather experimental evidence for the co-resonant signal enhancement, the bidirectional sensor was removed from the SEM/FIB instrument at various points of the resonance frequency adjustment process and transferred into the dSFM equipment to conduct the measurements presented in section 7.2.2.

7.2.1 Resonance frequency of the iron-filled carbon nanotube

It was observed that the resonance frequency of FeCNTs after the resonance frequency adjustment process decreased considerably when being removed from the SEM equipment for some time, as was required for the dSFM experiments. Furthermore, it was also found, that the resonance frequency of the FeCNT decreased continually when the resonance frequency-adjusted sensors were stored under ambient conditions for longer periods of time,

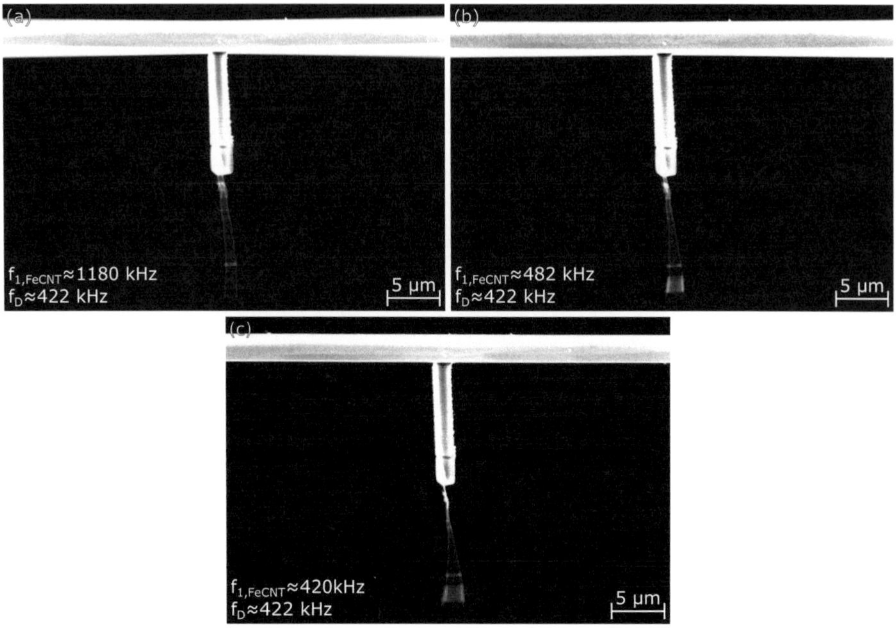

Figure 7.3: Collective oscillation of the cantilever and the carbon nanotube. The system was excited close to the resonance frequency of the second flexural mode of the cantilever. (a) Ridgid behavior of the FeCNT, (b) collective resonant oscillation of cantilever and FeCNT, (c) to protect the FeCNT the excitation amplitude was greatly reduced. In this case the oscillation of the cantilever was no longer visible.

but hardly when stored under vacuum. Although the exact origin of this behavior was not examined, it is reasonable to assume that the reason for this reduction of the resonance frequency is given by an increase of the mass of the freshly deposited materials by reaction with or inclusion of molecules or atoms contained in the atmosphere. Corresponding observations for different electron deposited materials have been reported and discussed in literature [229, 230]. A transfer through air could not be avoided with the available equipment and a determination of the resonance frequency of the FeCNT in the dSFM equipment was not possible. This resulted in a comparatively large uncertainty of the measured resonance frequencies of the FeCNT when used in calculations regarding the dSFM measurements. Because of this, the resonance frequencies in the following were calculated as mean values from SEM based vibration measurements before and after the dSFM measurements. The corresponding margins of errors indicate the whole range in which the resonance frequency drifted.

Because of the residual gas in the SEM system, especially from the deposition process, the resonance frequency of the FeCNT still changed during SEM imaging even if the precursor gas injection was turned off. To minimize the corresponding drifts, the values of the resonance frequency of the FeCNT during and after the adjustment process were determined by manually searching the excitation frequency corresponding to the highest FeCNT oscillation amplitude. This method is much faster than measuring amplitude versus driving frequency values for a resonance curve fit. Of course this faster method includes an increased uncertainty of the measurement resonance frequency but due to the even larger uncertainty induced by the drift when exposed to air, this fast determination method was deemed sufficient.

During the experiments it had to be considered that the tungsten needle that was used to detune the cantilever could detach due to instabilities in the equipment. In such an event an excitation with a large amplitude and a frequency close to that of the resonance frequency of the cantilever could destroy the sensor. Therefore, when searching for flexural resonances of the FeCNT close to the resonance frequency of the cantilever only small maximal excitation amplitudes were applied to the vibration stage.

A cylindrical nanotube or nanowire with a perfect circular cross section has no distinguished transverse direction resulting in a degenerate flexural resonance mode with the oscillation direction determined by the applied mechanical excitation only. In a real system, however, this symmetry can be broken by small variations in the geometry or mechanical parameters of the nanotube or its attachment as for example an asymmetry of the cross section or a variation of the Young's modulus. This results in two transverse flexural modes with very similar resonance frequencies. Corresponding observations have been described for several types of nanowires with circular or other highly symmetric cross sections in literature [231, 232, 233].

In the above discussed case of the unmodified carbon nanotube, no second first-order flexural resonance was found, even when observing the FeCNT from different directions. This changed, however, during the resonance frequency adjustment process. Then two flexural modes were often found, possibly due to asymmetries induced by the deposited material [232].

Such an occurrence of several modes with very similar resonance frequencies introduces additional levels of complexity into the system, as the oscillation mode of the cantilever can couple to all of these resonance modes. Completely describing such a complex system would need more complex models or even different methods as finite element simulations, which is beyond the scope of this work. Furthermore, it has to be mentioned that the amplitude observed in SEM is only a projection on the image plane and resonances might have larger amplitudes in a different direction. Material deposition can rotate the orientation of the oscillation planes of these flexural modes. Therefore, it is possible that a second oscillation mode cannot be reliably detected for certain frequency adjustment states as it oscillates perpendicular to the SEM image plane [232].

7.2.2 Bidirectional scanning force microscopy

The bidirectional sensor with the adjusted FeCNT resonance frequency was employed to measure the magnetostatic force gradients between the iron filling of the FeCNT and the Co/Pt multilayer test sample already used in the experiments in chapter 5. Measurements for both sensitivity directions were performed for various frequency matching states, starting from the unmodified FeCNT case up to states with very small resonance frequency separations of the sensor's subsystems. For the lateral measurements only the visible main peak of the amplitude versus driving frequency curve of coupled system was employed, as the second peak could not be detected by the system due to its small maximal amplitude. For all these measurements the frequency modulation method was used and the size of the scan area was given by 5 µm × 5 µm at resolution of 512 × 512 data points. In order to obtain a clear signal the scan speeds and phase-locked loop parameters were chosen for each frequency shift data map acquisition according to the maximal magnitude of the measured signal. These adjustments to the measurement parameter were necessary as the measured maximal signals heights covered several order of magnitude depending on the resonance frequency matching state of the sensor and the measured sensitivity direction. Furthermore, for low signal heights and corresponding long measurement durations, the data was corrected by a plane subtraction to compensate for thermal drift of the resonance frequency of the sensor. All experiments were conducted in high vacuum and at room temperature.

The amplitude in the perpendicular operation mode was chosen to be 10 nm. In the lateral operation mode small amplitudes of ≈ 4 nm at the free end of the cantilever were used in order to prevent damage and nonlinear behavior of the CNT due to the expected amplitude amplification. This amplitude corresponded to a lateral amplitude of the free end of the FeCNT of ≈ 0.3 nm, not considering the expected co-resonant amplitude amplification. Furthermore, the mean tip-sample separation was not determined for the measurements in order to protect the adjusted FeCNT, as this would have included contact with the sample. Instead, a vertical sensor position was chosen at which both operation directions show a reasonable frequency shift signal. This was especially crucial for very close frequency matching states, as the co-resonant enhanced lateral operation mode signal was observed to easily exceed 100 Hz, which is the limit for the dSFM equipment. The frequency shift maps for both operation directions were then acquired at the same vertical sensor position. If the signal levels required longer measurement duration, this could have resulted in a slightly different measurement height of the frequency shift maps due to a thermal drift of the mean tip-sample separation. However, the absolute error was expected to be small and was not considered for the evaluations. In the following an exemplary evaluation is demonstrated for one of these data sets.

7.2.3 Experimental amplification factor

For this examplary evaluation an intermediate resonance frequency adjustment state was chosen at which the adjusted resonance frequency of the first flexural oscillation mode of the FeCNT is given by $f^{*,adj}_{1,FeCNT} = (426 \pm 5)$ kHz. The corresponding obtained perpendicular and lateral sensitivity direction frequency shift maps are shown in figures 7.4(a) and (b).

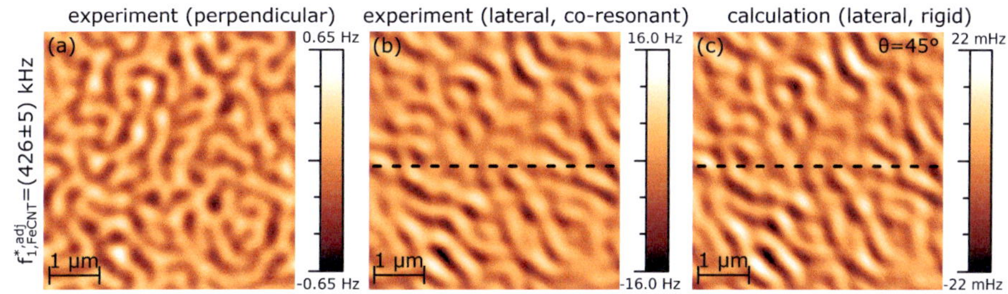

Figure 7.4: Frequency shift data acquired with the bidirectional sensor with co-resonant enhancement and calculation for a non-enhanced sensor: (a) experimental perpendicular operation mode data, (b) experimental enhanced lateral operation mode data, (c) data calculated based on (a) for a non-enhanced sensor. The dashed line indicates the section shown in figure 7.7. The rotation angle used for the calculation is $\theta = 45°$.

As demonstrated in chapter 5 a lateral magnetostatic force gradient measurement on the Co/Pt test sample shows the strongest contrast at sample positions close to domain boundaries that are oriented perpendicular to the oscillation direction of the interaction tip. In the experimental data shown in figure 7.4(b), which was recorded with the enhanced sensor's lateral operation mode, the strongest contrast was found close to domain borders that exhibited an angle of 45° with respect to the expected direction, however. This indicates that the FeCNT's oscillation direction deviated from the oscillation direction of the sensor's spacer element by this angle.

A superposition of such nonaligned oscillations should in principle result in a complex combined trajectory of the FeCNT's free end. The excited amplitude of the spacer element was small, however, and the FeCNT's amplitude much larger due to the co-resonant amplitude amplification. Therefore, this trajectory can still be considered approximately line shaped along the oscillation direction of the nanotube. Furthermore, the frequency shift signal of a co-resonantly modified sensor is dominated by the oscillation of the carbon nanotube. Any components that arise from a contribution of the cantilever's oscillation direction are very small due to its high effective spring constant in the lateral operation mode and can be neglected.

Therefore, the sensitivity direction is assumed to be simply rotated by an angle of $\theta = 45°$. A sketch clarifying the oscillation directions and the definition of the angle can be found in figure 7.5. The influence of this rotation angle on the signal strength is discussed below in section 7.2.5.

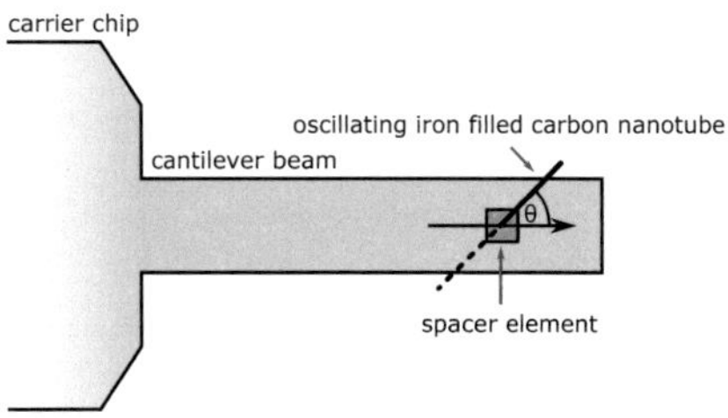

Figure 7.5: Sketch of the bidirectional sensor with a misaligned co-resonant oscillation of the iron-filled carbon nanotube to illustrate the definition of the rotation angle θ. The arrows indicate the oscillation direction of the spacer element's free end that is connected to the FeCNT. The orientation of the sketch corresponds to the sensor orientation in the measurements to acquire the data shown in figures 7.4 and 7.9.

The aim of this chapter is to demonstrate the co-resonant signal amplification. To this end an experimental signal amplification factor has to be derived. The measured resonance frequency of the first flexural mode of the cantilever was found to be far lower than even the adjusted resonance frequency of the first flexural mode of the FeCNT. Therefore, the influence of the frequency adjustment on the perpendicular operation mode signal is negligible according to the co-resonant sensor concept. Hence, it is possible to recalculate experimental perpendicular frequency shift data into corresponding force gradient data according to equation 5.8 with the factor $-2k^{dyn}_{x_{tip},1}/f^*_1 = (-9.54 \pm 0.91) \cdot 10^{-5}$ Ns/m.

As was demonstrated in chapter 5 the lateral force gradients can be calculated for a rigid FeCNT according to equation 5.13 based on two-dimensional maps of perpendicular data. From these calculated force gradients, a hypothetical frequency shift signal expected for a rigid FeCNT can be obtained by using equation 5.9. In the corresponding calculations that employed the same software as before a rotation of the sensitivity direction by 45°

was included and the hypothetical frequency shift signal was calculated from the calculated lateral force gradient data with the corresponding factor $-2k^{dyn}_{\Delta x,2}/f^*_2 = (-1.60 \pm 0.34) \cdot 10^{-3}\text{Ns/m}$. The calculated frequency shift map is show in figure 7.4(c). The general shape of the experimentally obtained and the calculated lateral frequency shift maps are in very good agreement and the co-resonant signal enhancement is already visible in the very different scales.

In order to obtain an experimental signal amplification factor the distributions of the experimental and calculated frequency shift values were examined by creating corresponding histograms. As is shown in figure 7.6 these distributions, which were derived from the measurements conducted on the Co/Pt test sample with high lateral symmetry, could be fitted well by Gaussian functions. The experimental amplification factor of the sensor is then defined as the ratio of the full widths at half maximum of the fits to the experimental and the theoretical lateral data distributions as these can be used to describe the data range. This definition assumes that the frequency shift signal is much higher than the noise levels, which was the case for the discussed data set. This resulted in an experimental amplification factor of:

$$\Lambda^{exp} = (6.2 \pm 2.1) \cdot 10^2. \tag{7.5}$$

The margins of error of this experimental amplification factor mainly originated from that of the calculated frequency shift signal, which in turn primarily stemmed from the error of the spring constant of the lateral operation mode of the sensor.

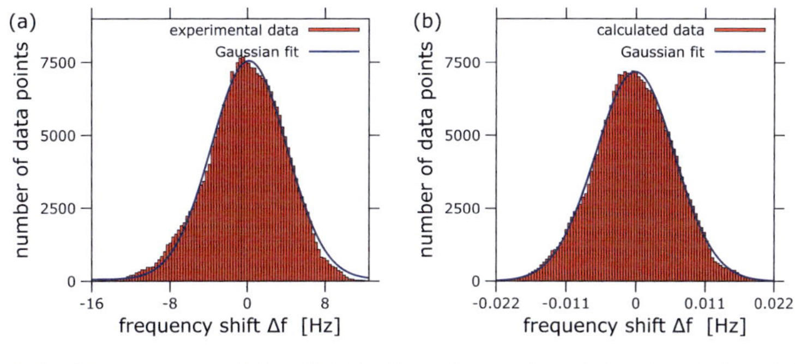

Figure 7.6: Histogramms of the distribution of experimental and calculated frequency shift data: (a) experimental co-resonant lateral operation mode data and (b) calculated data for the case of a rigid iron-filled carbon nanotube. The Gaussian fits are indicated.

The good agreement between the calculated data multiplied by this amplification factor and the experimental data shown in figure 7.7, demonstrates that this experimental amplification factor describes the signal enhancement well. This high amplification factor validates the signal amplification due to the resonance frequency matching of cantilever and nanotube according to the co-resonant sensor concept.

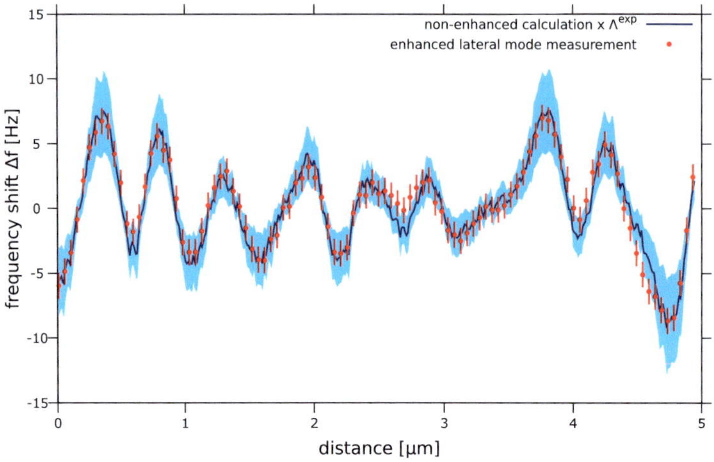

Figure 7.7: Line section of the co-resonantly enhanced lateral operation mode frequency shift data. The data obtained with the co-resonant sensor is compared to the data calculated for the sensor with a rigid FeCNT multiplied by the obtained experimental amplification factor. The light blue area indicates the margins of error for the calculated data. The error on the amplification factor was not included in these limits, as this would account for the errors of the calculated data twice. Only every 5th experimental data point is shown for better visibility.

There is a slight notable asymmetry in the histogram of the experimental data. This can be explained by a different amplification for negative and positive force gradients as was already shown in figure 6.4 of the discussion of the co-resonant sensor concept. Such a asymmetry arises if the measured force gradients are not by several orders of magnitudes lower than the spring constant of the smaller oscillator and correspond to a non-linear behavior of the sensor. The involved magnetostatic force gradients in the discussed case were already determined during the derivation of the calculated frequency shift map and were found to be in the range of $\approx \pm 35\ \mu\text{N/m}$. This is well within the range at which such an notable asymmetry can occur according to the coupled harmonic oscillator model. This asymmetry could have been prevented if a higher vertical sensor position, corresponding to smaller force gradients had been chosen for the experiments. This would also have further lowered the signal in the perpendicular operation mode, however, resulting in greater difficulties to measure it. Furthermore, this asymmetry only notably affects the extremal values of the experimental data and therefore only slightly the width of a corresponding

Gaussian fit. As additionally the derived experimental amplification factor described the data well, the asymmetry was neglected for these evaluations. It should be kept in mind for practical applications of the co-resonantly enhanced sensors, however.

Besides the experimental verification of the co-resonant signal enhancement, the demonstrated method to determine an experimental amplification factor constitutes a practical method to calibrate a bidirectional FeCNT-based sensor with co-resonant signal enhancement independent of the knowledge of the dynamic properties of the carbon nanotube.

7.2.4 Comparison to theoretical amplifications factors

Experimental amplification factors were determined for various frequency adjustment states. The obtained flexural resonance frequencies of the FeCNT and the corresponding experimental amplification factors are shown in figure 7.8. The theoretical amplification factors in this figure were calculated for the simple coupled harmonic oscillator model as well as aligned FeCNT and spacer element oscillations according to equation 6.8. These calculations were based on the obtained values for the respective effective spring constants and resonance frequencies of cantilever and iron-filled carbon nanotube, respectively. Furthermore, it was assumed that the spring constant of the carbon nanotube was not altered by the material deposition at its free end and that the resonance frequency of the cantilever subsystem was constant throughout the experiments. Please note that in the derived margins of error for the amplification factors in figure 7.8 the uncertainties of the resonance frequencies of the FeCNT are not included, as they are indicated separately.

In the experiments the unmodified sensor was first used to acquire the dSFM data shown in figures 7.9(a) and (b). The resonance frequency was determined as $f^{*,adj}_{1,FeCNT} = (1178 \pm 2)$ kHz. Please note that a resonance frequency drift was observed despite no material was intentional deposited before the dSFM measurement. However, the measurements discussed in section 7.1.1 took quite some time during which an unintentional material deposition is probable. The lateral operation mode signal exhibited very high measurement noise mainly due to the combination of a small signal frequency shift signal and the low oscillation amplitude. For this state no rotation of the oscillation direction was found and the derived small experimental amplification factor $\Lambda_{exp} = 1.44 \pm 0.49$ was, within its margin of error, in agreement with both the theoretical calculation and the completely rigid case with no amplification. This is the same situation as was discussed for the standard bidirectional sensors in section 5.5.6.

Following the first large frequency adjustment, two first-order flexural resonance modes of the FeCNT were found in the vibration experiments. The sensor stayed functional

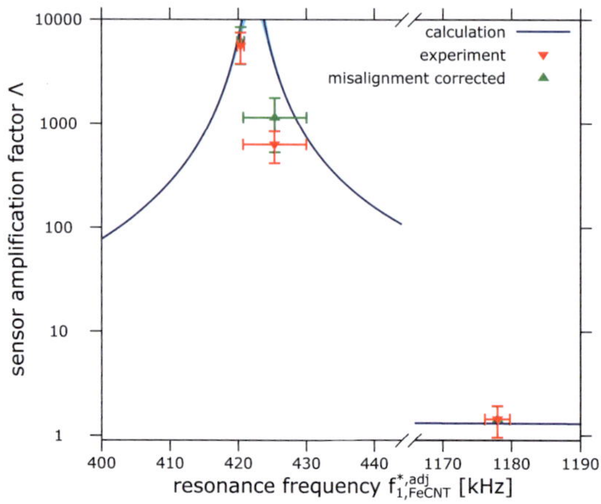

Figure 7.8: Sensor amplification factor Λ versus adjusted resonance frequency $f^{*,adj}_{1,FeCNT}$ of the iron-filled carbon nanotube. Please note that the calculated curve is not a fit to the data but was derived for an aligned FeCNT oscillation ($\theta = 0°$) according to equation 6.8 based on the resonance frequencies and spring constants of the subsystems.

during further intermediate frequency matching steps and exhibited signal amplification in corresponding dSFM experiments. However, it was unclear to which of the oscillation modes these amplifications could be attributed to and a measurement of the resonance frequencies of the coupled system indicated that the simple harmonic oscillator model could not be used to describe the sensor's behavior in the these states as discussed above. Therefore, these frequency adjustment states were not included in the evaluation shown in figure 7.8.

After further material depositions, the sensor reached the frequency matching state already used for the exemplary determination of the experimental amplification factor above. In this case only one flexural resonance of the FeCNT at a resonance frequency of $f^{*,adj}_{1,FeCNT} = (426 \pm 5)$ kHz was found in the vicinity of the resonance frequency of the cantilever at $f^*_2 = (422.13 \pm 0.01)$ kHz. Furthermore, the measured resonance frequencies of the coupled system were in agreement with a simple coupled oscillator model with two masses. Since the resonance frequency separation was very small it was assumed that the observed flexural resonance of the FeCNT dictated the sensor's overall behavior.

The final frequency adjustment step was special in the sense that no new material was deposited and only the resonance frequency drift due to exposure to atmosphere was used for the adjustment. This resulted in a comparatively small resonance frequency drift making this the most precise measurement in terms of the uncertainty of the resonance frequency

$f^{*,adj}_{1,FeCNT} = (420 \pm 1)$ kHz of the FeCNT. In this case a measurement in the dSFM equipment revealed a notable change of the resonance frequency of the main mode of the sensor of about +280 Hz when compared to the previous measurements. This change is in accordance with the behavior of a coupled oscillator system with two masses as the resonance frequency of the corresponding FeCNT resonance mode in this matching state was lower than that of cantilever. This resulted in the coupled mode with the higher resonance frequency being the main mode of the sensor. An analysis of the corresponding dSFM data presented in figures 7.9(d) and (e) resulted in an experimental amplification factor of $\Lambda_{exp} = (5.6 \pm 1.9) \cdot 10^3$ in very good agreement with the theoretically expected value. The contrast features of the experimental lateral frequency shift map looked more blurred than those of the calculated lateral frequency shift map. The reason for this could simply be an unexpected high drift towards a higher vertical sensor position between the corresponding perpendicular and lateral measurements, which would correspond to an even higher amplification factor. This could also indicate a loss of resolution due to the co-resonant amplitude amplification of the FeCNT or an artifact due to operating the dSFM equipment at unusually high frequency shift signals, however, and should be studied in future experiments.

7.2.5 Angle correction of the amplification factor

The amplification factors shown in figure 7.8 agree with the theoretical expectation within their margins of error. The amplification factor at the resonance frequency $f^{*,adj}_{1,FeCNT} = (426 \pm 5)$ kHz is still far lower than the theoretical expectation, however, and is only in agreement with the calculated value almost at the limit of its corresponding maximum errors. Even though this is possible, it is quite unlikely; instead, this may have have been caused by the large angle of the misaligned oscillation direction of the FeCNT.

To consider the corresponding influence on the measured frequency shifts, again the simple driven coupled harmonic oscillator model was used. To derive the resonance frequencies of the combined system, they were again approximated by the eigenfrequencies of a corresponding undamped system.

In such a system the non-aligned oscillations can be described by two different effects. On the one hand, the projection of the amplitude of the lateral oscillation of the bidirectional sensor on the oscillation direction of the FeCNT that drives its oscillation is reduced by a factor of $\cos(\theta)$, with $\theta \in (-90°, 90°)$ being the angle between the oscillation directions and $\theta = 0$ describing an aligned oscillation. On the other hand, the force exerted by the FeCNT on the sensor along the oscillation direction of the spacer element is also reduced

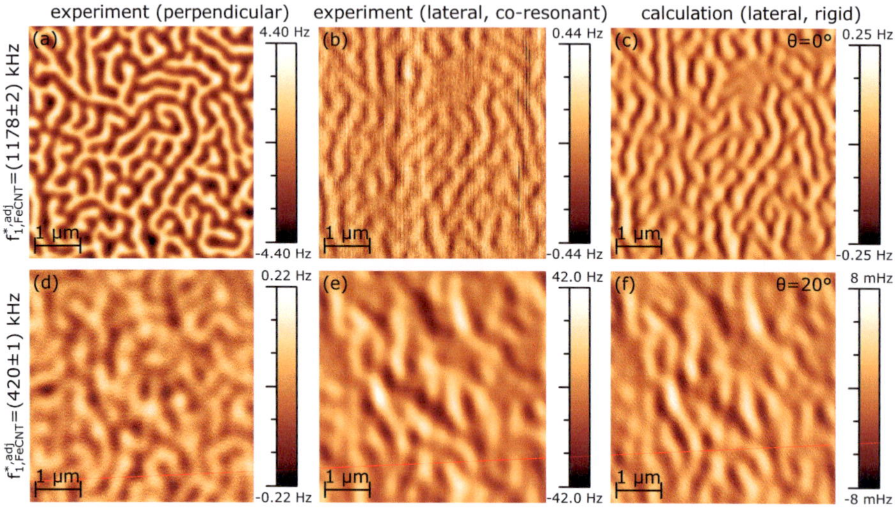

Figure 7.9: Additional frequency shift data sets experimentally obtained with the co-resonant sensor and corresponding calculated data that was used for the evaluation of the amplification factor in figure 7.8. The rotation angles used for the calculations as well as the resonance frequency adjustment states of the FeCNT that this data was attributed to are indicated. As indicated (a)-(c) correspond to the unmodified sensor with $f^{*,adj}_{1,FeCNT} = (1179.71 \pm 0.25)$ kHz and (d)-(f) to a very well matched state with $f^{*,adj}_{1,FeCNT} = (420 \pm 1)$ kHz.

by a factor of $\cos(\theta)$. Such a system can be described by a system of differential equations in analogy to equation 6.3 with the matrices

$$\mathbf{M} = \begin{pmatrix} m_1 & 0 \\ 0 & m_2 \end{pmatrix}, \quad \mathbf{D} = \begin{pmatrix} 0 & 0 \\ 0 & 0 \end{pmatrix}, \quad \mathbf{K} = \begin{pmatrix} k_1 + \cos(\theta)k_2 & -\cos(\theta)k_2 \\ -\cos(\theta)k_2 & k_2 + k_3 \end{pmatrix} \tag{7.6}$$

and $\vec{F}(t) = (0,0)$. By using the ansatz $\vec{s}(t) = \vec{\hat{A}} \cdot cos(\omega t)$ [27, pp. 212f.], this results in an eigenvalue problem with the eigenfrequencies as solutions. These were found to be:

$$\omega^2(\theta, k_3)_{a/b} = \frac{T}{2} \mp \sqrt{\left(\frac{T}{2}\right)^2 - \frac{\omega_1^2\omega_2^2}{k_1}\left(k_1 + \cos(\theta)k_2 + \frac{k_1 k_3}{k_2} + \cos(\theta)k_3 - \cos^2(\theta)k_2\right)} \tag{7.7}$$

with

$$T := \omega_2^2\left(1 + \frac{k_3}{k_2}\right) + \omega_1^2\left(1 + \cos(\theta)\frac{k_2}{k_1}\right). \tag{7.8}$$

To derive these equations $m_i = k_i/\omega_i^2$ was used. These eigenfrequencies can be employed to calculate the factor

$$\gamma_{rot,\theta}(k_3) = \frac{\Delta\omega_{a/b}(\theta, k_3)}{\Delta\omega_{a/b}(0, k_3)} = \frac{\omega_{a/b}(\theta, k_3) - \omega_{a/b}(\theta, 0)}{\omega_{a/b}(0, k_3) - \omega_{a/b}(0, 0)} \tag{7.9}$$

which describes the relative reduction of the frequency shift signal of the sensor with respect to the force gradient due to the non-aligned oscillation directions. Equation 7.9 was examined by numerical evaluations and it was found that for small k_3 the factor $\gamma_{rot,\theta}$ only depends on the angle θ and decreases with increasing angle. At very large misalignment angles close to 90° this factor was found to be close to zero. This means that the contribution of an oscillation mode with such a strong misalignment to the frequency shift signal of the sensor is extremely small.

Corresponding factors for small k_3 were calculated for all experimental resonance frequency adjustment states by assuming that the resonance frequency adjustment does not influence the spring constant of the FeCNT. The specific eigenfrequency used for the calculation depended on the peak of the coupled system which was measured. These factors were used to correct the experimental amplification factors assuming a maximum error of ±5° on the determination of the angle and the results are shown in figure 7.8. With this correction the agreement between the theoretically expected and the experimentally obtained sensor amplification factors was improved for all data sets with a non vanishing oscillation direction misalignment. This further indicates the applicability of the simple coupled harmonic oscillator model with two masses to describe the sensor's behavior. It also shows that the amplification factors calculated based on equation 6.8 can be used to describe the sensors signal enhancement.

7.2.6 Discussion of the results

The reasonable agreement between the experimental results and the theory shown above are an indication that the equation 6.8 for the amplification factor derived in the previous chapter can be used to at least estimate the signal amplification by the co-resonant resonance frequency matching. This is further supported by the very good agreement for the resonance frequency matching state corresponding to a FeCNT resonance frequency of $f_{1,FeCNT}^{*,adj} = (420 \pm 1)$ kHz. In this case separation between the resonance frequencies of the subsystems is very small and possible competing influence of a nearby second resonance of the FeCNT should be negligible.

These results cannot be considered a solid proof of the applicability of equation 6.8 to calculate the amplification factor, however, due to the large uncertainties of the resonance frequency of the FeCNT at the time of the dSFM measurement, the low number of valid data points and the difficulty to account for the second first-order flexural resonance of the FeCNT. Future experiments on this matter should include the ability to reliably and quickly rotate the sensor along the axis of the FeCNT within the SEM equipment to measure the amplitude of the resonances from different angles, so no resonances are hidden in order to obtain more valid data points.

Nevertheless, the observation of a much higher frequency shift signal in the lateral than in the perpendicular mode for the states with closely matched resonance frequencies indicates, without any assumptions, the signal enhancement by matching the resonance frequency of a first-order flexural mode of the FeCNT to that of the cantilever's second flexural mode.

Finally, the observation of a maximal experimental amplification factor of $\Lambda_{exp} = (5.6 \pm 1.9) \cdot 10^3$ affirms the improvement of the bidirectional sensors by the co-resonant resonance frequency matching and the viability of the co-resonant sensor concept.

7.3 Applicability and remaining challenges

With the experiments presented in this chapter, it was demonstrated that FeCNT-based bidirectional sensors with co-resonant signal enhancement exhibit a very high frequency shift signal. A quantitative evaluation of the experimental data is still possible with the knowledge of the amplification factor Λ of the sensor or, an equivalent description, its effective co-resonant spring constant $k_\Lambda = k^{dyn}_{\Delta x,2}/\Lambda$. This amplification factor can either be derived by calibration measurements or estimated from the effective spring constants and the corresponding resonance frequencies of the subsystems. As it was discussed above, however, obtaining the correct resonance frequency for the FeCNT can be a difficult task and the influence of a second first-order flexural mode of the FeCNT with a similar resonance frequency is still an open question.

For the application of the bidirectional sensors with co-resonant signal enhancement, the issue of the misalignment of the oscillation directions has to be addressed. The rotation of the oscillation plane of the FeCNT can be explained by an asymmetry of the CNTs geometry as discussed above. Such an asymmetry can change due to the material deposition explaining the different oscillation angles of the matching states [232]. An easy solution would be to use electron beam induced material deposition and removal techniques to intentionally create a geometrical asymmetry of the FeCNT that results in a first-order flexural mode with a well defined oscillation direction aligned to that of the cantilever. Additionally, this would also separate the resonance frequencies of the two first-order flexural modes of the FeCNT, thus ensuring that the resulting sensor's behavior can be well described by the simple coupled oscillator model with two masses and facilitating the determination of the correct resonance frequency during the adjustment process. Such a modification influences the spring constant of the FeCNT, however, and the viability of such an approach has to be explored in the future. A further challenge for a practical application of these sensors is the drift of the FeCNT's resonance frequency when exposed to ambient conditions. A solution may be provided by using only a Pt precursor gas when adjusting the resonance frequency. The deposited material is less prone to oxidation than pure carbon due to the platinum content, although it has to be kept in mind that it still contains a large fraction of carbon. This method has already been used to produce further cantilever sensors with FeCNTs and co-resonant signal enhancement and the frequency drift was observed to be far less severe [234]. The asymmetry in the amplification for positive and negative force gradient of the sensors has to be kept in mind for practical purposes. Therefore, these sensors should only be employed for measurements of sufficiently small force gradients.

A practical question concerning any new dynamic mode scanning force microscopy sensor concept is that of the ultimate limit of the detectable force gradient. For a cantilever force sensor described by a harmonic oscillator model and a laser beam-deflection based measurement [235] the expression for this minimal detectable force gradient is based on the discussion of the mean square deflection of the cantilever due to the thermal energy $k_B T$ and results in the equation [8, pp.92-95]

$$\left(\frac{\partial F_s}{\partial s}\right)_{min} = \sqrt{\frac{4}{3} k_B T} \sqrt{\frac{2B_\omega}{\pi^3 A^2}} \sqrt{\frac{k_i}{\omega_i Q_i}}, \tag{7.10}$$

which is also valid for higher oscillation modes. Here, k_i, ω_i, Q_i are the spring constant, resonance frequency and Q factor of the cantilever's mode i respectively, A the cantilever's amplitude and B_ω is the bandwidth of the measurement. While equation 7.10 is also applicable for individual FeCNTs used directly as force sensors, it cannot easily be adapted for the coupled system. To derive a corresponding equation for the co-resonant sensors, the mean square deflection of the cantilever due to the combined thermal vibrations of the asymmetrical coupled system has to be obtained. All required parameters of the coupled system can be derived from obtained dynamical properties, i.e. the resonance frequencies, effective spring constants and Q factors of the subsystems as summarized in tables 7.1 and 7.2. Even then this is not an easy task [236], however, and therefore the minimal detectable force gradient of the co-resonant bidirectional sensors remains an open question for future work. Additionally, the quantitative influence of the co-resonant amplitude amplification on the lateral resolution of the measurement is an topic that is also of interest for practical applications and should be addressed.

Finally, it is emphasized that the application of the co-resonant sensor concept for scanning force microscopy sensors is not limited to the bidirectional sensors used above. In principle, standard perpendicular sensitivity dSFM sensors could also be equipped with a frequency adjusted CNT with its long axis parallel to the cantilever. However, the low bending stiffness of the CNT would make it prone to the 'jump-to-contact' instability during measurements. In contrast, the co-resonant sensor concept can directly be applied to the torsional resonance mode. This would be even more advantageous as the resonance frequency of the first torsional resonance frequency of the cantilever is higher than that of the flexural mode. Due to the comparatively small resonance frequency separation even an unmodified FeCNT can lead to a notable signal enhancement for the FeCNT-cantilever combination used in the experiments above.

8 Conclusion and outlook

The main topic of this thesis is the advancement, development and implementation of novel sensor concepts for dynamic mode scanning force microscopy. First of all, a recently introduced sensor concept for bidirectional scanning force microscopy was explored in terms of its applicability to quantitative force gradient measurements in two different sensitivity directions, lateral and perpendicular to a sample surface. Required expressions for the dynamic spring constants for the lateral and the perpendicular operation mode were derived based on the Euler-Bernoulli beam theory as well as on geometrical and energy considerations. The validity of these expressions was experimentally demonstrated by using a corresponding sensor with an iron-filled carbon nanotube interaction tip to measure magnetostatic interactions close to the surface of a magnetic multilayer sample. A good quantitative agreement was found both between the acquired experimental data and calculated results, as well as between the effective magnetic monopole approximation parameters of the tip obtained from a scanning force microscopy measurement and independent geometric data. This agreement validates the correctness of the derived expressions and the viability of the sensors for quantitative measurements in the perpendicular and a lateral direction. By evaluating a torsional resonance mode measurement performed with these novel sensors it was also demonstrated that tridirectional force gradient measurements are possible.

To solve the challenge of comparatively high spring constants in the lateral operation mode, a new co-resonant sensor concept was proposed to combine the ease-of-detection of microscaled oscillators with the high sensitivity of low spring constant oscillating nanosystems. This concept is based on an asymmetric coupled harmonic oscillator model: By matching the resonance frequencies of both oscillator subsystems it is possible to greatly enhance the frequency shift signal of the main resonance mode measured on the bigger oscillator system in response to an external force gradient acting on the smaller oscillator. This signal enhancement can be described in terms of a signal amplification factor for which an approximate analytical expression was derived in this thesis. Although this sensor concept was first proposed to improve the bidirectional sensors examined in the first part of this work, it is of a very general nature and may be applied to many different kinds of mechanical oscillation based sensors with respect to the detection of mass changes and external force gradients.

This co-resonant sensor concept was then implemented by reducing the resonance frequency of the iron-filled carbon nanotube interaction tip of a bidirectional sensor by electron beam induced mass deposition, until it was similar to the resonance frequency corresponding to the lateral operation mode of the sensor. The thus modified sensor was then used to measure magnetostatic force gradient data in both operation directions on a magnetic multilayer sample. The resulting experimental data was compared with theoretical calculations for a non-modified sensor, validating a signal amplification of more than three orders of magnitude. The signal amplification factor of a such a co-resonant sensor can either be obtained by a calibration measurement or it can be calculated from the effective spring constants and resonance frequencies of cantilever and iron-filled carbon nanotube. Therefore, these kind of sensors are still applicable to quantitative scanning force microscopy measurements. To the knowledge of the author this was also the first demonstration of scanning force microscopy based on the superior mechanical properties of a carbon nanotube resonator.

In addition to these results, a further application of the iron-filled carbon nanotubes was also explored: they were used as a well-known and easy to handle test system for nano-SQUID magnetometers, demonstrating their ability to examine magnetic nanoparticles.

The development of the co-resonant sensor concept is not concluded. Some questions, which are important for practical applications remain, such as the influence of the sensor's functionality based on the coupled system on the possible measurement speed as well as the minimal detectable force gradient. Furthermore, the model for the co-resonant signal enhancement should be expanded to account for the observed multiple flexural resonances with similar resonance frequencies. Additionally, a further optimization and improvement of the fabrication process of these sensors is desirable. This should also include an improvement of the vibration stage-based resonance frequency determination method to allow for an easy examination from different observation angles.

Now that these new quantitative scanning force microscopy methods have been explored and developed, they can be employed to measure the properties of interesting sample systems. Several experiments are in progress to study free standing tubes of rolled up magnetic thin films or exchange biased ion irradiation patterned magnetic samples. Furthermore, the study of other interesting systems such as magnetic skyrmions would also be a fascinating task for these methods.

Finally, experiments on co-resonantly enhanced cantilever magnetometry sensors with unprecedented sensitivity, another application of the co-resonant sensor concept, have already progressed very far [237] and further results will be published in the near future. Together with the co-resonantly enhanced bidirectional sensors, these implementations with the well-known iron-filled carbon nanotubes only provide a first glimpse of the great potential co-resonant sensor concept. Future implementations of co-resonant sensor designs, based on even smaller resonators such as single-walled carbon nanotubes promise even higher sensitivities for studying extremely small force gradients or mass changes.

Appendices

A Analytical solutions of the coupled harmonic oscillator

The steady state solutions for a generic coupled harmonic oscillator with two masses as defined by the system of differential equations 6.3 were solved according to the method discussed in section 6.2.1. The amplitudes $\vec{\hat{A}} = (\hat{A}_1, \hat{A}_2)$ of the complex ansatz for the deflection $\vec{s}(t) = \vec{\hat{A}} \cdot \exp(i\omega_D t)$ that resulted from the solution of the system of linear equations were found to be

$$\hat{A}_1 = \frac{F_0}{\chi^2 + \kappa^2}(\upsilon + i\nu) \tag{A.1}$$

$$\hat{A}_2 = \frac{F_0}{\chi^2 + \kappa^2}(\epsilon + i\zeta) \tag{A.2}$$

with the definitions

$$\chi := k_{12}k_{23} - k_2^2 + \omega_D^4 m_1 m_2 - \omega_D^2 \left(m_1 k_{23} + m_2 k_{12} + d_{12}d_{23} - d_2^2\right) \tag{A.3}$$

$$\kappa := -\omega_D^3 \left(m_1 d_{23} + m_2 d_{12}\right) + \omega_D \left(d_{12}k_{23} + d_{23}k_{12} - 2d_2 k_2\right) \tag{A.4}$$

$$\upsilon := \left(-\omega_D^2 m_2 + k_{23}\right)\chi + \omega_D d_{23}\kappa \tag{A.5}$$

$$\nu := \left(\omega_D^2 m_2 - k_{23}\right)\kappa + \omega_D d_{23}\chi \tag{A.6}$$

$$\epsilon := \omega_D d_2 \kappa + k_2 \chi \tag{A.7}$$

$$\zeta := \omega_D d_2 \chi - k_2 \kappa \tag{A.8}$$

$$k_{ij} := k_i + k_j$$

$$d_{ij} := d_i + d_j. \tag{A.9}$$

With the magnitudes

$$A_1 = \frac{F_0}{\chi^2 + \kappa^2}\sqrt{\upsilon^2 + \nu^2} \tag{A.10}$$

$$A_2 = \frac{F_0}{\chi^2 + \kappa^2}\sqrt{\epsilon^2 + \zeta^2} \tag{A.11}$$

and arguments

$$\phi_1 = \arg(\upsilon + i\nu) \tag{A.12}$$

$$\phi_2 = \arg(\epsilon + i\zeta) \tag{A.13}$$

of these complex amplitudes and using Euler's formula they can be written as

$$\hat{A}_1 = A_1(\cos(\phi_1) + i\sin\phi_1) = A_1 \exp i\phi_1 \tag{A.14}$$

$$\hat{A}_2 = A_2(\cos(\phi_2) + i\sin\phi_2) = A_2 \exp i\phi_2. \tag{A.15}$$

The solutions to the equations of motion are then given by the real parts of the complex deflections

$$s_1(t) = \Re(\hat{A}_1 \exp(i\omega_D t)) = \Re(A_1 \exp(i\omega_D t + \phi_1)) = A_1 \cos(\omega_D t + \phi_1) \tag{A.16}$$

$$s_2(t) = \Re(\hat{A}_2 \exp(i\omega_D t)) = \Re(A_2 \exp(i\omega_D t + \phi_2)) = A_2 \cos(\omega_D t + \phi_2). \tag{A.17}$$

From these equations it is obvious that the magnitudes A_1, A_2 and arguments ϕ_1, ϕ_2 of the complex amplitudes are also the desired amplitudes and phases of the real solution. The argument function $\arg z$ of a complex number z can be defined as [238, pp.34f.]:

$$\arg(z) = \arg(x + iy) = \begin{cases} \arctan\left(\frac{y}{x}\right) & \text{if } x > 0 \\ \arctan\left(\frac{y}{x}\right) + \pi & \text{if } x < 0 \text{ and } y \geq 0 \\ \arctan\left(\frac{y}{x}\right) - \pi & \text{if } x < 0 \text{ and } y < 0 \\ +\frac{\pi}{2} & \text{if } x = 0 \text{ and } y > 0 \\ -\frac{\pi}{2} & \text{if } x = 0 \text{ and } y < 0 \\ \text{undetermined} & \text{if } x = 0 \text{ and } y = 0 \end{cases} \tag{A.18}$$

Furthermore, the calculated phase ϕ_2 can have a positive value. This case is unphysical, as the reaction would be prior to the cause. Therefore, the phase has to be corrected by subtraction of 2π as the derived solution is invariant against such an operation.

B Approximation for the frequency shift signal of a coupled oscillator

In order to derive an analytical approximation for the frequency shift signal of the coupled harmonic oscillator equation 6.5 for the eigenfrequencies of the undamped system was used as a starting point:

$$\omega^2_{a/b,k_3} = \frac{\omega^2_{12} + \omega^2_{23}}{2} \mp \sqrt{\frac{\left(\omega^2_{12} - \omega^2_{23}\right)^2}{4} + \frac{k_2^2}{m_1 m_2}}.$$

With the ratio of the eigenfrequency of the smaller system to the first combined eigenfreqency $\omega_2 = \alpha\omega_{12}$ as well as the equations for the combined eigenfrequencies $\omega^2_{ij} = (k_i + k_j)/m_i$ and the eigenfrequencies of the subsystems $\omega^2_i = k_i/m_i$, the individual parts of this formula were rewritten:

$$\frac{\omega^2_{12} + \omega^2_{23}}{2} = \frac{\omega^2_{12}}{2}\left(1 + \alpha^2 + \alpha^2\frac{k_3}{k_2}\right) \tag{B.1}$$

$$\frac{k_2^2}{m_1 m_2} = \frac{\omega^4_{12}}{4} 4\alpha^2 \frac{k_2}{k_1 + k_2} \tag{B.2}$$

$$\frac{\left(\omega^2_{12} - \omega^2_{23}\right)^2}{4} = \frac{\omega^4_{12}}{4}\left(1 - \alpha^2 - \alpha^2\frac{k_3}{k_2}\right)^2 \tag{B.3}$$

$$= \frac{\omega^4_{12}}{4}\left(1 - 2\alpha^2 + \alpha^4\left(1 + \left(\frac{k_3}{k_2}\right)^2\right) + \left(2\alpha^4 - 2\alpha^2\right)\frac{k_3}{k_2}\right) \tag{B.4}$$

$$\approx \frac{\omega^4_{12}}{4}\left(1 - 2\alpha^2 + \alpha^4 + \left(2\alpha^4 - 2\alpha^2\right)\frac{k_3}{k_2}\right). \tag{B.5}$$

Assuming $k_2 >> k_3$, the approximation $(1 + (k_3/k_2)^2) \approx 1$ was used for the last equation. These equations were employed to derive an approximation for the eigenfrequencies from equation 6.5

$$\omega^2_{a/b,k_3} \approx \frac{\omega^2_{12}}{2}\left(1 + \alpha^2 + \alpha^2\frac{k_3}{k_2} \mp \zeta\sqrt{1 + \frac{2\alpha^4 - 2\alpha^2}{\zeta^2}\frac{k_3}{k_2}}\right) \tag{B.6}$$

with the definition

$$\zeta^2 := 1 + 2\alpha^2\left(\frac{2k_2}{k_1 + k_2} - 1\right) + \alpha^4.$$

from equation 6.9. To expand the square root the corresponding Taylor series of $\sqrt{1+x}$ was used [238, p.1011]:

$$\sqrt{1+x} = 1 + \frac{1}{2}x - \frac{1}{8}x^2 + \mathcal{O}\left(x^3\right). \tag{B.7}$$

This series converges for $1 \geq |x|$. In addition to this criteria for convergence the linear term alone is a only a good approximation, if its absolute value is much larger than the following square term:

$$\frac{|x|}{2} >> \frac{|x^2|}{8} \tag{B.8}$$

or for simplification neglecting the factor 4

$$1 >> |x|\,. \tag{B.9}$$

This expansion was applied to equation B.6 yielding

$$\omega^2_{a/b,k_3} = \frac{\omega^2_{12}}{2}\left(1+\alpha^2+\alpha^2\frac{k_3}{k_2} \mp \zeta\left(1+\frac{2\alpha^4-2\alpha^2}{2\zeta^2}\frac{k_3}{k_2} - \mathcal{O}\left(\left(\frac{2\alpha^4-2\alpha^2}{2\zeta^2}\frac{k_3}{k_2}\right)^2\right)\right)\right). \tag{B.10}$$

The higher order terms can be omitted if according to equation B.9

$$\left|\frac{\zeta^2}{2\alpha^4-2\alpha^2}\right| >> \left|\frac{k_3}{k_2}\right| \tag{B.11}$$

holds. To obtain the frequency shift, the eigenfrequencies and not their squares were needed. Therefore the square root of equation B.10

$$\omega_{a/b,k_3} \approx \frac{\omega_{12}}{\sqrt{2}}\sqrt{1+\alpha^2 \mp \zeta + \left(\alpha^2 \mp \frac{2\alpha^4-2\alpha^2}{2\zeta}\right)\frac{k_3}{k_2}} \tag{B.12}$$

without the higher order terms was considered. Expanding the square root resulted in

$$\omega_{a/b,k_3} \approx \frac{\omega_{12}}{\sqrt{2}}\left(\sqrt{1+\alpha^2\mp\zeta} + \frac{\alpha^2 \mp \frac{2\alpha^4-2\alpha^2}{2\zeta}}{2\sqrt{1+\alpha^2\mp\zeta}}\frac{k_3}{k_2} + \mathcal{O}\left(\left(\frac{\alpha^2 \mp \frac{2\alpha^4-2\alpha^2}{2\zeta}}{\sqrt{1+\alpha^2\mp\zeta}}\frac{k_3}{k_2}\right)^2\right)\right). \tag{B.13}$$

Here, the higher orders can be neglected if

$$\left|\frac{\sqrt{1+\alpha^2\mp\zeta}}{\alpha^2 \mp \frac{2\alpha^4-2\alpha^2}{2\zeta}}\right| >> \left|\frac{k_3}{k_2}\right| \tag{B.14}$$

is satisfied. With this, a linear expression for the approximation of the frequency shifts in response to an external force gradient $k_3 = \Delta k$ was formulated:

$$\Delta\omega_{a/b} = \omega_{a/b,\Delta k} - \omega_{a/b,0} \approx \frac{\omega_{12}}{\sqrt{2}}\frac{\alpha^2 \mp \frac{\alpha^4-\alpha^2}{\zeta}}{2\sqrt{1+\alpha^2\mp\zeta}}\frac{\Delta k}{k_2}. \tag{B.15}$$

Writing this in analogy to the frequency shift of a single harmonic oscillator and using equation B.13 this finally resulted in the frequency shift in response to an external force gradient as given by equation 6.7:

$$\Delta\omega_{a/b} \approx \frac{\omega_{a/b,0}\Lambda_{a/b}}{2k_1}\Delta k$$

with the amplification factor from equation 6.8

$$\Lambda_{a/b} = \frac{k_1}{k_2}\frac{\alpha^2 \mp \frac{\alpha^4-\alpha^2}{\zeta}}{1+\alpha^2 \mp \zeta}$$

The quality of the approximation may be derived by studying the expressions B.11 and B.14. As these expressions are complicated they did not allow for a derivation of a simple, generally valid result. However, for a given set of parameters k_1, k_2, α they can easily be evaluated analytically to check the scope of validity of the approximation. A numerical study showed that for the parameters used in chapter 6 these expressions were easily satisfied for $k_3 << k_2$ as long as α was not extremely close to 1.

Bibliography

[1] G. Binnig, H. Rohrer, Ch. Gerber, and E. Weibel. Surface studies by scanning tunneling microscopy. *Phys. Rev. Lett.*, 49(1):57–61, 1982.

[2] G. K. Binnig. Atomic force microscope and method for imaging surfaces with atomic resolution, 1988. US Patent 4,724,318.

[3] G. Binnig, C. F. Quate, and Ch. Gerber. Atomic force microscope. *Phys. Rev. Lett.*, 56(9):930–933, 1986.

[4] T. Mühl and J. Körner. Verfahren und Vorrichtung zur Messung von Kraftgradienten bei der Kraftmikroskopie. Patent application. 10/13/2011.

[5] F. Wolny, T. Mühl, U. Weissker, K. Lipert, J. Schumann, A. Leonhardt, and B. Büchner. Iron filled carbon nanotubes as novel monopole-like sensors for quantitative magnetic force microscopy. *Nanotechnology*, 21(43):435501, 2010.

[6] S. Philippi, U. Weißker, T. Mühl, A. Leonhardt, and B. Büchner. Room temperature magnetometry of an individual iron filled carbon nanotube acting as nanocantilever. *J. Appl. Phys.*, 110(8):084319, 2011.

[7] S. Morita, R. Wiesendanger, and E. Meyer, editors. *Noncontact atomic force microscopy.* Nanoscience and technology. Springer, Berlin, Heidelberg, 2002.

[8] E. Meyer, H. J. Hug, and R. Bennewitz. *Scanning probe microscopy.* Springer, Berlin, Heidelberg, 2004.

[9] F. J. Giessibl. Forces and frequency shifts in atomic-resolution dynamic-force microscopy. *Phys. Rev. B*, 56(24):16010–16015, 1997.

[10] H.-U. Danzebrink, L. Koenders, G. Wilkening, A. Yacoot, and H. Kunzmann. Advances in scanning force microscopy for dimensional metrology. *CIRP Ann. Manuf. Techn.*, 55(2):841–878, 2006.

[11] B. Voigtländer. *Scanning Probe Microscopy*. Nanoscience and technology. Springer, Berlin, Heidelberg, 2015.

[12] Q. Zhong, D. Inniss, K. Kjoller, and V.B. Elings. Fractured polymer/silica fiber surface studied by tapping mode atomic force microscopy. *Surf. Sci. Lett.*, 290(1-2):L688–L692, 1993.

[13] R. García and A. San Paulo. Amplitude curves and operating regimes in dynamic atomic force microscopy. *Ultramicroscopy*, 82(1-4):79–83, 2000.

[14] F. J. Giessibl. Atomic resolution of the silicon (111)-(7x7) surface by atomic force microscopy. *Science*, 267(5194):68–71, 1995.

[15] H. Ueyama, M. Ohta, Y. Sugawara, and S. Morita. Atomically resolved InP(110) surface observed with noncontact ultrahigh vacuum atomic force microscope. *Jpn. J. Appl. Phys.*, 34(8B):L1086–L1088, 1995.

[16] Y. Sugawara, M. Ohta, H. Ueyama, and S. Morita. Defect motion on an InP(110) surface observed with noncontact atomic force microscopy. *Science*, 270(5242):1646–1648, 1995.

[17] F. J. Giessibl, S. Hembacher, H. Bielefeldt, and J. Mannhart. Subatomic features on the silicon (111)-(7x7) surface observed by atomic force microscopy. *Science*, 289(5478):422–425, 2000.

[18] D. Anselmetti, R. Lüthi, E. Meyer, T. Richmond, M. Dreier, J. E. Frommer, and H.-J. Güntherodt. Attractive-mode imaging of biological materials with dynamic force microscopy. *Nanotechnology*, 5(2):87–94, 1994.

[19] L. Gross, F. Mohn, N. Moll, P. Liljeroth, and G. Meyer. The chemical structure of a molecule resolved by atomic force microscopy. *Science*, 325(5944):1110–1114, 2009.

[20] F. J. Giessibl. High-speed force sensor for force microscopy and profilometry utilizing a quartz tuning fork. *Appl. Phys. Lett.*, 73(26):3956, 1998.

[21] T. R. Albrecht, P. Grütter, D. Horne, and D. Rugar. Frequency modulation detection using high-q cantilevers for enhanced force microscope sensitivity. *J. Appl. Phys.*, 69(2):668–673, 1991.

[22] O. Wolter, T. Bayer, and J. Greschner. Micromachined silicon sensors for scanning force microscopy. *J. Vac. Sci. Technol. B*, 9(2):1353–1357, 1991.

[23] F. J. Giessibl. Advances in atomic force microscopy. *Rev. Mod. Phys.*, 75(3):949–983, 2003.

[24] S. Rast, C. Wattinger, U. Gysin, and E. Meyer. Dynamics of damped cantilevers. *Rev. Sci. Instrum.*, 71(7):2772, 2000.

[25] D. Sarid. *Scanning force microscopy.* Number 2 in Oxford series on optical sciences. Oxford University Press, New York, Oxford, 1991.

[26] M. A. Lantz, H. J. Hug, P. J. A. van Schendel, R. Hoffmann, S. Martin, A. Baratoff, A. Abdurixit, H.-J. Güntherodt, and Ch. Gerber. Low temperature scanning force microscopy of the Si (111)-(7x7) surface. *Phys. Rev. Lett.*, 84(12):2642–2645, 2000.

[27] W. Nolting. *Grundkurs Theoretische Physik 1: Klassische Mechanik*, volume 1 of *Grundkurs Theoretische Physik*. Springer, Berlin, Heidelberg, New York, 6th revised edition, 2002.

[28] L. D. Landau and E. M. Lifschitz. *Lehrbuch der theoretischen Physik: Mechanik.* Akademie - Verlag, Berlin, 9th revised edition, 1976.

[29] S.N. Magonov, V. Elings, and M.-H. Whangbo. Phase imaging and stiffness in tapping-mode atomic force microscopy. *Surf. Sci.*, 375(2-3):L385–L391, 1997.

[30] U. Hartmann. Magnetic force microscopy. *Annu. Rev. Mater. Sci.*, 29(1):53–87, 1999.

[31] P. A. Tipler and G. Mosca. *Physik.* Springer, 6th corrected edition, 2012.

[32] R. García and R. Pérez. Dynamic atomic force microscopy methods. *Surf. Sci. Rep.*, 47(6-8):197–301, 2002.

[33] Y. Martin and H. K. Wickramasinghe. Magnetic imaging by "force microscopy" with 1000 ˚a resolution. *Appl. Phys. Lett.*, 50(20):1455, 1987.

[34] H. Hölscher and U. D. Schwarz. Theory of amplitude modulation atomic force microscopy with and without Q-Control. *Int. J. Nonlinear Mech.*, 42(4):608–625, 2007.

[35] F. J. Giessibl. A direct method to calculate tip-sample forces from frequency shifts in frequency-modulation atomic force microscopy. *Appl. Phys. Lett.*, 78(1):123–125, 2001.

[36] J. E. Sader and S. P. Jarvis. Accurate formulas for interaction force and energy in frequency modulation force spectroscopy. *Appl. Phys. Lett.*, 84(10):1801, 2004.

[37] M. Ternes, C. P. Lutz, C. F. Hirjibehedin, F. J. Giessibl, and A. J. Heinrich. The force needed to move an atom on a surface. *Science*, 319(5866):1066–1069, 2008.

[38] D. Rugar, H. J. Mamin, P. Guethner, S. E. Lambert, J. E. Stern, I. McFadyen, and T. Yogi. Magnetic force microscopy: General principles and application to longitudinal recording media. *J. Appl. Phys.*, 68(3):1169, 1990.

[39] L. Meirovitch. *Elements of vibration analysis.* McGraw-Hill, New York, 1975.

[40] W. T. Thomson. *Theory of vibration with applications.* Chapmann & Hall, London, Glasgow, New York, 4th edition, 1993.

[41] S. Timoshenko. *Vibration Problems in Engineering.* D. Van Nostrand Company, Inc., Princeton, NJ, 3rd revised edition, 1956.

[42] A. Raman, J. Melcher, and R. Tung. Cantilever dynamics in atomic force microscopy. *Nano Today*, 3(1-2):20–27, 2008.

[43] R. C. Hibbeler. *Mechanics of Materials.* Pearson Prentice Hall, Upper Saddle River, NJ, 6th edition, 2005.

[44] D. Rüdiger and A. Kneschke. *Technische Mechanik*, volume 2. B.G. Teubner, Leipzig, 1962.

[45] G. Haehner. Dynamic spring constants for higher flexural modes of cantilever plates with applications to atomic force microscopy. *Ultramicroscopy*, 110(7):801–806, 2010.

[46] J. P. Cleveland, S. Manne, D. Bocek, and P. K. Hansma. A nondestructive method for determining the spring constant of cantilevers for scanning force microscopy. *Rev. Sci. Instrum.*, 64(2):403, 1993.

[47] U. Rabe, K. Janser, and W. Arnold. Vibrations of free and surface-coupled atomic force microscope cantilevers: Theory and experiment. *Rev. Sci. Instrum.*, 67(9):3281, 1996.

[48] J. C. Bruch and T. P. Mitchell. Vibrations of a mass-loaded clamped-free timoshenko beam. *J. Sound Vib.*, 114(2):341–345, 1987.

[49] R. W. Stark and W. M. Heckl. Higher harmonics imaging in tapping-mode atomic-force microscopy. *Rev. Sci. Instrum.*, 74(12):5111, 2003.

[50] F. Zypman. Internal damping for noncontact atomic force microscopy cantilevers. *J. Vac. Sci. Technol. B*, 28(3):C4E24, 2010.

[51] R. W. Stark and W. M. Heckl. Fourier transformed atomic force microscopy: tapping mode atomic force microscopy beyond the Hookian approximation. *Surf. Sci.*, 457(1-2):219–228, 2000.

[52] M. A. Poggi, A. W. McFarland, J. S. Colton, and L. A. Bottomley. A method for calculating the spring constant of atomic force microscopy cantilevers with a nonrectangular cross section. *Anal. Chem.*, 77(4):1192–1195, 2005.

[53] J. L. Hutter and J. Bechhoefer. Calibration of atomic-force microscope tips. *Rev. Sci. Instrum.*, 64(7):1868, 1993.

[54] S. M. Cook, T. E. Schäffer, K. M. Chynoweth, M. Wigton, R. W. Simmonds, and K. M. Lang. Practical implementation of dynamic methods for measuring atomic force microscope cantilever spring constants. *Nanotechnology*, 17(9):2135–2145, 2006.

[55] J. E. Sader, J. W. M. Chon, and P. Mulvaney. Calibration of rectangular atomic force microscope cantilevers. *Rev. Sci. Instrum.*, 70(10):3967, 1999.

[56] R. S. Gates and M. G. Reitsma. Precise atomic force microscope cantilever spring constant calibration using a reference cantilever array. *Rev. Sci. Instrum.*, 78(8):086101, 2007.

[57] J. E. Sader, I. Larson, P. Mulvaney, and L. R. White. Method for the calibration of atomic force microscope cantilevers. *Rev. Sci. Instrum.*, 66(7):3789, 1995.

[58] J. E. Sader and L. White. Theoretical analysis of the static deflection of plates for atomic force microscope applications. *J. Appl. Phys.*, 74(1):1, 1993.

[59] E. W. Wong. Nanobeam mechanics: Elasticity, strength, and toughness of nanorods and nanotubes. *Science*, 277(5334):1971–1975, 1997.

[60] J. J. Sáenz, N. García, P. Grütter, E. Meyer, H. Heinzelmann, R. Wiesendanger, L. Rosenthaler, H. R. Hidber, and H.-J. Güntherodt. Observation of magnetic forces by the atomic force microscope. *J. Appl. Phys.*, 62(10):4293, 1987.

[61] T. Göddenhenrich, H. Lemke, U. Hartmann, and C. Heiden. Magnetic force microscopy of domain wall stray fields on single-crystal iron whiskers. *Appl. Phys. Lett.*, 56(25):2578, 1990.

[62] A. Moser, H. J. Hug, I. Parashikov, B. Stiefel, O. Fritz, H. Thomas, A. Baratoff, H.-J. Güntherodt, and P. Chaudhari. Observation of single vortices condensed into a

vortex-glass phase by magnetic force microscopy. *Phys. Rev. Lett.*, 74(10):1847–1850, 1995.

[63] P. Milde, D. Kohler, J. Seidel, L. M. Eng, A. Bauer, A. Chacon, J. Kindervater, S. Muhlbauer, C. Pfleiderer, S. Buhrandt, C. Schütte, and A. Rosch. Unwinding of a skyrmion lattice by magnetic monopoles. *Science*, 340(6136):1076–1080, 2013.

[64] J. D. Jackson. *Klassische Elektrodynamik*. de Gruyter, Berlin, Boston, 5th edition, 2014.

[65] C. D. Wright and E. W. Hill. Reciprocity in magnetic force microscopy. *Appl. Phys. Lett.*, 67(3):433, 1995.

[66] H. J. Hug, B. Stiefel, P. J. A. van Schendel, A. Moser, R. Hofer, S. Martin, H.-J. Güntherodt, S. Porthun, L. Abelmann, J. C. Lodder, G. Bochi, and R. C. O'Handley. Quantitative magnetic force microscopy on perpendicularly magnetized samples. *J. Appl. Phys.*, 83(11):5609, 1998.

[67] A. Hubert and R. Schäfer, editors. *Magnetic domains*. Springer, Berlin, Heidelberg, New York, 1st corrected edition, 1998.

[68] U. Hartmann, T. Göddenhenrich, and C. Heiden. Magnetic force microscopy: Current status and future trends. *J. Magn. Magn. Mater.*, 101(1-3):263–270, 1991.

[69] U. Hartmann. The point dipole approximation in magnetic force microscopy. *Phys. Lett. A*, 137(9):475–478, 1989.

[70] U. Hartmann. Fundamentals and special applications of non-contact scanning force microscopy. *Adv. Electron. Electron Phys.*, 87:49–200, 1993.

[71] J. Lohau, S. Kirsch, A. Carl, G. Dumpich, and E. F. Wassermann. Quantitative determination of effective dipole and monopole moments of magnetic force microscopy tips. *J. Appl. Phys.*, 86(6):3410, 1999.

[72] Th. Kebe and A. Carl. Calibration of magnetic force microscopy tips by using nanoscale current-carrying parallel wires. *J. Appl. Phys.*, 95(3):775, 2004.

[73] J. W. Li, J. P. Cleveland, and R. Proksch. Bimodal magnetic force microscopy: Separation of short and long range forces. *Appl. Phys. Lett.*, 94(16):163118, 2009.

[74] J. Schwenk, M. Marioni, S. Romer, N. R. Joshi, and H. J. Hug. Non-contact bimodal magnetic force microscopy. *Appl. Phys. Lett.*, 104(11):112412, 2014.

[75] NanoScan AG. High-resolution magnetic force microscope hr-MFM. Promotional Brochure, 2007.

[76] NanoScan AG. High-resolution magnetic force microscope. User Manual version 2.0, 2009.

[77] H.-J. Butt, B. Cappella, and M. Kappl. Force measurements with the atomic force microscope: Technique, interpretation and applications. *Surf. Sci. Rep.*, 59(1-6):1–152, 2005.

[78] G. Meyer and N. M. Amer. Optical-beam-deflection atomic force microscopy: The NaCl (001) surface. *Appl. Phys. Lett.*, 56(21):2100, 1990.

[79] C. A. J. Putman, B. G. De Grooth, N. F. Van Hulst, and J. Greve. A detailed analysis of the optical beam deflection technique for use in atomic force microscopy. *J. Appl. Phys.*, 72(1):6, 1992.

[80] Timothy Ashworth. Personal communication: 7/25/2012. NanoScan AG.

[81] U. Dürig. Relations between interaction force and frequency shift in large-amplitude dynamic force microscopy. *Appl. Phys. Lett.*, 75(3):433–435, 1999.

[82] B. Gotsmann, C. Seidel, B. Anczykowski, and H. Fuchs. Conservative and dissipative tip-sample interaction forces probed with dynamic AFM. *Phys. Rev. B*, 60(15):11051–11061, 1999.

[83] P. F. Schmidt, L. J. Balk, R. Balschke, W. Bröcker, E. Demm, L. Engel, R. Göcke, H. Hantsche, R. Hauert, E.R. Krefting, Th. Müller, H. Raith, M. Roth, and J. Woodtli. *Praxis der Raserelektronenmikroskopie und Mikrobereichsanalyse*, volume 444: Metechnik of *Kontakt und Studium*. expert, Renningen-Malmsheim, 1994.

[84] T. E. Everhart and R. F. M. Thornley. Wide-band detector for micro-microampere low-energy electron currents. *J. Sci. Instrum.*, 37(7):246–248, 1960.

[85] J. Goldstein, D. E. Newbury, P. Echlin, D. C. Joy, A. D. Romig, C. E. Lyman, C. Fiori, and E. Lifshin. *Scanning Electron Microscopy and X-Ray Microanalysis*. Plenum Press, New York, London, 2nd edition, 1992.

[86] H. Seiler. Secondary electron emission in the scanning electron microscope. *J. Appl. Phys.*, 54(11):R1, 1983.

[87] D. C. Joy and J. B. Pawley. High-resolution scanning electron microscopy. *Ultramicroscopy*, 47(1-3):80–100, 1992.

[88] Siegfried Menzel. Personal comunication: 6/28/2013. Relaying information from Carl Zeiss AG service technician.

[89] L. A. Giannuzzi and F. A. Stevie, editors. *Introduction to Focused Ion Beams*. Springer Science+Buissness Media, 1st edition, 2005.

[90] T. D. Yuzvinsky, A. M. Fennimore, W. Mickelson, C. Esquivias, and A. Zettl. Precision cutting of nanotubes with a low-energy electron beam. *Appl. Phys. Lett.*, 86(5):053109, 2005.

[91] F. Wolny, U. Weissker, T. Mühl, A. Leonhardt, S. Menzel, A. Winkler, and B. Büchner. Iron-filled carbon nanotubes as probes for magnetic force microscopy. *J. Appl. Phys.*, 104(6):064908, 2008.

[92] S. Philippi. *The Mechanical Response of Individual Ferromagnetic Nanowires in Applied Fields*. PhD thesis, Technische Universität Dresden, 2013.

[93] U. Weißker. *Synthesis and mechanical properties of iron-filled carbon nanotubes*. PhD thesis, Technische Universität Dresden, 2013.

[94] F. Wolny. *Magnetic properties of individual iron filled carbon nanotubes and their application as probes for magnetic force microscopy*. PhD thesis, Technische Universität Dresden, 2011.

[95] A. Krüeger. *Carbon Materials and Nanotechnology*. WILEY-VCH, Weinheim, 2010.

[96] M. D. Hanwell, D. E. Curtis, D. C. Lonie, T. Vandermeersch, E. Zurek, and G. R. Hutchison. Avogadro: an advanced semantic chemical editor, visualization, and analysis platform. *J. Cheminf.*, 4(1):17, 2012.

[97] M. S. Dresselhaus, G. Dresselhaus, and R. Saito. Physics of carbon nanotubes. *Carbon*, 33(7):883–891, 1995.

[98] S. Iijima. Helical microtubules of graphitic carbon. *Nature*, 354(6348):56–58, 1991.

[99] D. H. Robertson, D. W. Brenner, and J. W. Mintmire. Energetics of nanoscale graphitic tubules. *Phys. Rev. B*, 45(21):12592–12595, 1992.

[100] G. Overney, W. Zhong, and D. Tománek. Structural rigidity and low frequency vibrational modes of long carbon tubules. *Z. Phys. D Atom. Mol. Clust.*, 27(1), 1993.

[101] J.-P. Salvetat, J.-M. Bonard, N. H. Thomson, A. J. Kulik, L. Forr, W. Benoit, and L. Zuppiroli. Mechanical properties of carbon nanotubes. *Appl. Phys. A Mater. Sci. Process.*, 69(3):255–260, 1999.

[102] C. Li and T.-W. Chou. Elastic moduli of multi-walled carbon nanotubes and the effect of van der waals forces. *Compos. Sci. Technol.*, 63(11):1517–1524, 2003.

[103] M. M. J. Treacy, T. W. Ebbesen, and J. M. Gibson. Exceptionally high young's modulus observed for individual carbon nanotubes. *Nature*, 381(6584):678–680, 1996.

[104] M. Yu. Strength and breaking mechanism of multiwalled carbon nanotubes under tensile load. *Science*, 287(5453):637–640, 2000.

[105] J.-C. Charlier, X. Gonze, and J.-P. Michenaud. First-principles study of the electronic properties of graphite. *Phys. Rev. B*, 43(6):4579–4589, 1991.

[106] N. Hamada, S. Sawada, and A. Oshiyama. New one-dimensional conductors: Graphitic microtubules. *Phys. Rev. Lett.*, 68(10):1579–1581, 1992.

[107] T. W. Ebbesen. Carbon nanotubes. *Annu. Rev. Mater. Sci.*, 24(1):235–264, 1994.

[108] J. W. Mintmire and C. T. White. Electronic and structural properties of carbon nanotubes. *Carbon*, 33(7):89–902, 1995.

[109] J.-C. Charlier, X. Blase, and S. Roche. Electronic and transport properties of nanotubes. *Rev. Mod. Phys.*, 79(2):677–732, 2007.

[110] E. T. Thostenson, Z. Ren, and T.-W. Chou. Advances in the science and technology of carbon nanotubes and their composites: a review. *Compos. Sci. Technol.*, 61(13):1899–1912, 2001.

[111] V. Popov. Carbon nanotubes: properties and application. *Mater. Sci. Eng. R*, 43(3):61–102, 2004.

[112] A. Leonhardt, S. Hampel, C. Müller, I. Mönch, R. Koseva, M. Ritschel, D. Elefant, K. Biedermann, and B. Büchner. Synthesis, properties, and applications of ferromagnetic-filled carbon nanotubes. *Chem. Vapor Depos.*, 12(6):380–387, 2006.

[113] C. Müller, D. Golberg, A. Leonhardt, S. Hampel, and B. Büchner. Growth studies, TEM and XRD investigations of iron-filled carbon nanotubes. *Phys. Stat. Sol. A*, 203(6):1064–1068, 2006.

[114] S. Hampel, A. Leonhardt, D. Selbmann, K. Biedermann, D. Elefant, Ch. Müller, T. Gemming, and B. Büchner. Growth and characterization of filled carbon nanotubes with ferromagnetic properties. *Carbon*, 44(11):2316–2322, 2006.

[115] M. Löffler, U. Weissker, T. Mühl, T. Gemming, and B. Büchner. Robust determination of young's modulus of individual carbon nanotubes by quasi-static interaction with Lorentz forces. *Ultramicroscopy*, 111(2):155–158, 2011.

[116] L. S. Dubrovinsky, N. A. Dubrovinskaia, S. K. Saxena, S. Rekhi, and T. LeBihan. Aggregate shear moduli of iron up to 90 GPa and 1100 K. *J. Alloy. Comp.*, 297(1-2):156–161, 2000.

[117] K. Lipert, S. Bahr, F. Wolny, P. Atkinson, U. Weißker, T. Mühl, O. G. Schmidt, B. Büchner, and R. Klingeler. An individual iron nanowire-filled carbon nanotube probed by micro-Hall magnetometry. *Appl. Phys. Lett.*, 97(21):212503, 2010.

[118] P. Banerjee, F. Wolny, D. V. Pelekhov, M. R. Herman, K. C. Fong, U. Weissker, T. Mühl, Y. Obukhov, A. Leonhardt, B. Büchner, and Hammel. C. P. Magnetization reversal in an individual 25 nm iron-filled carbon nanotube. *Appl. Phys. Lett.*, 96(25):252505, 2010.

[119] J. Crangle and G. M. Goodman. The magnetization of pure iron and nickel. *Proc. Roy. Soc. Lond. A*, 321(1547):477–491, 1971.

[120] D.C. Jiles. *Introduction to Magnetism and Magnetic Materials.* Taylor & Francis, 2nd edition, 1998.

[121] M. U. Lutz, U. Weissker, F. Wolny, C. Müller, M. Löffler, T. Mühl, A. Leonhardt, B. Büchner, and R. Klingeler. Magnetic properties of α-Fe and Fe_3C nanowires. *J. Phys. Conf. Ser.*, 200(7):072062, 2010.

[122] A. Aharoni. Angular dependence of nucleation by curling in a prolate spheroid. *J. Appl. Phys.*, 82(3):1281, 1997.

[123] S. Blundell. *Magnetism in condensed matter.* Oxford master series in condensed matter physics. Oxford, Oxford, New York, 1st reprinted edition, 2001.

[124] H. Dai, J. H. Hafner, A. G. Rinzler, D. T. Colbert, and R. E. Smalley. Nanotubes as nanoprobes in scanning probe microscopy. *Nature*, 384(6605):147–150, 1996.

[125] N. R. Wilson and J. V. Macpherson. Carbon nanotube tips for atomic force microscopy. *Nat. Nanotechnol.*, 4(8):483–491, 2009.

[126] C. V. Nguyen, K.-J. Chao, R. M. D. Stevens, L. Delzeit, A. Cassell, J. Han, and M. Meyyappan. Carbon nanotube tip probes: stability and lateral resolution in scanning probe microscopy and application to surface science in semiconductors. *Nanotechnology*, 12(3):363–367, 2001.

[127] E. S. Snow, P. M. Campbell, and J. P. Novak. Single-wall carbon nanotube atomic force microscope probes. *Appl. Phys. Lett.*, 80(11):2002, 2002.

[128] M. C. Strus, A. Raman, C.-S. Han, and C. V. Nguyen. Imaging artefacts in atomic force microscopy with carbon nanotube tips. *Nanotechnology*, 16(11):2482–2492, 2005.

[129] H. Kuramochi, T. Uzumaki, M. Yasutake, A. Tanaka, H. Akinaga, and H. Yokoyama. A magnetic force microscope using CoFe-coated carbon nanotube probes. *Nanotechnology*, 16(1):24–27, 2004.

[130] N. R. Wilson and J. V. Macpherson. Single-walled carbon nanotubes as templates for nanowire conducting probes. *Nano Lett.*, 3(10):1365–1369, 2003.

[131] S. Porthun, L. Abelmann, S. J. L. Vellekoop, J. C. Lodder, and H. J. Hug. Optimization of lateral resolution in magnetic force microscopy. *Appl. Phys. A Mater. Sci. Process.*, 66(7):S1185–S1189, 1998.

[132] S. Vock, F. Wolny, T. Mühl, R. Kaltofen, L. Schultz, B. Büchner, C. Hassel, J. Lindner, and V. Neu. Monopole-like probes for quantitative magnetic force microscopy: Calibration and application. *Appl. Phys. Lett.*, 97(25):252505, 2010.

[133] B. C. Stipe, H. J. Mamin, T. D. Stowe, T. W. Kenny, and D. Rugar. Magnetic dissipation and fluctuations in individual nanomagnets measured by ultrasensitive cantilever magnetometry. *Phys. Rev. Lett.*, 86(13):2874–2877, 2001.

[134] D. P. Weber, D. Rüffer, A. Buchter, F. Xue, E. Russo-Averchi, R. Huber, P. Berberich, J. Arbiol, A. Fontcuberta i Morral, D. Grundler, and M. Poggio. Cantilever magnetometry of individual Ni nanotubes. *Nano Lett.*, 12(12):6139–6144, 2012.

[135] T. Schwarz, R. Wölbing, C. F. Reiche, B. Müller, M. J. Martínez-Pérez, T. Mühl, B. Büchner, R. Kleiner, and D. Koelle. Low-noise $YBa_2Cu_3O_7$ nano-SQUIDs for performing magnetization-reversal measurements on magnetic nanoparticles. *Phys. Rev. Appl.*, 3(4), 2015.

[136] W. Buckel and R. Kleiner. *Supraleitung*. WILEY-VCH, 6th revised edition, 2004.

[137] D. Dimos, P. Chaudhari, and J. Mannhart. Superconducting transport properties of grain boundaries in $YBa_2Cu_3O_7$ bicrystals. *Phys. Rev. B*, 41(7):4038–4049, 1990.

[138] M. Tinkham. *Introduction to superconductivity*. Dover publications, 2nd edition, 2004.

[139] J. Clarke and A. I. Braginski, editors. *The SQUID Handbook*, volume 1. WILEY-VCH, Weinheim, 2004.

[140] W. Wernsdorfer, K. Hasselbach, D. Mailly, B. Barbara, A. Benoit, L. Thomas, and G. Suran. DC-SQUID magnetization measurements of single magnetic particles. *J. Magn. Magn. Mater.*, 145(1-2):33–39, 1995.

[141] S. K. H. Lam and D. L. Tilbrook. Development of a niobium nanosuperconducting quantum interference device for the detection of small spin populations. *Appl. Phys. Lett.*, 82(7):1078, 2003.

[142] J. Nagel, K. B. Konovalenko, M. Kemmler, M. Turad, R. Werner, E. Kleisz, S. Menzel, R. Klingeler, B. Büchner, R. Kleiner, and D. Koelle. Resistively shunted $YBa_2Cu_3O_7$ grain boundary junctions and low-noise SQUIDs patterned by a focused ion beam down to 80 nm linewidth. *Supercond. Sci. Technol.*, 24(1):015015, 2011.

[143] R. Wölbing, T. Schwarz, B. Müller, J. Nagel, M. Kemmler, R. Kleiner, and D. Koelle. Optimizing the spin sensitivity of grain boundary junction nanoSQUIDs-towards detection of small spin systems with single-spin resolution. *Supercond. Sci. Technol.*, 27(12):125007, 2014.

[144] E. H. Frei, S. Shtrikman, and D. Treves. Critical size and nucleation field of ideal ferromagnetic particles. *Phys. Rev.*, 106(3):446–455, 1957.

[145] R. Skomski. Nanomagnetics. *J. Phys. Condens. Matter*, 15(20):R841–R896, 2003.

[146] G. S. Abo, Y.-K. Hong, J. Park, J. Lee, W. Lee, and B.-C. Choi. Definition of magnetic exchange length. *IEEE Trans. Magn.*, 49(8):4937–4939, 2013.

[147] J. Nagel, A. Buchter, F. Xue, O. F. Kieler, T. Weimann, J. Kohlmann, A. B. Zorin, D. Rüffer, E. Russo-Averchi, R. Huber, P. Berberich, A. Fontcuberta i Morral, D. Grundler, R. Kleiner, D. Koelle, M. Poggio, and M. Kemmler. Nanoscale multifunctional sensor formed by a Ni nanotube and a scanning Nb nanoSQUID. *Phys. Rev. B*, 88(6), 2013.

[148] T. Mühl, J. Körner, S. Philippi, C. F. Reiche, A. Leonhardt, and B. Büchner. Magnetic force microscopy sensors providing in-plane and perpendicular sensitivity. *Appl. Phys. Lett.*, 101(11):112401, 2012.

[149] U. Dürig. Interaction sensing in dynamic force microscopy. *New J. Phys.*, 2(1):5.1, 2000.

[150] S. Hembacher, F. J. Giessibl, and J. Mannhart. Force microscopy with light-atom probes. *Science*, 305(5682):380–383, 2004.

[151] O. Sahin, G. Yaralioglu, R. Grow, S. F. Zappe, A. Atalar, C. Quate, and O. Solgaard. High-resolution imaging of elastic properties using harmonic cantilevers. *Sens. Actuators A*, 114(23):183–190, 2004.

[152] S. Kawai, S. Kitamura, D. Kobayashi, S. Meguro, and H. Kawakatsu. An ultrasmall amplitude operation of dynamic force microscopy with second flexural mode. *Appl. Phys. Lett.*, 86(19):193107, 2005.

[153] T. R. Rodríguez and R. García. Compositional mapping of surfaces in atomic force microscopy by excitation of the second normal mode of the microcantilever. *Appl. Phys. Lett.*, 84(3):449–451, 2004.

[154] C. Dietz, M. Zerson, C. Riesch, A. M. Gigler, R. W. Stark, N. Rehse, and R. Magerle. Nanotomography with enhanced resolution using bimodal atomic force microscopy. *Appl. Phys. Lett.*, 92(14):143107, 2008.

[155] C. Dietz, E. T. Herruzo, J. R. Lozano, and R. Garcia. Nanomechanical coupling enables detection and imaging of 5 nm superparamagnetic particles in liquid. *Nanotechnology*, 22(12):125708, 2011.

[156] O. Pfeiffer, R. Bennewitz, A. Baratoff, E. Meyer, and P. Grütter. Lateral-force measurements in dynamic force microscopy. *Phys. Rev. B*, 65(16):161403, 2002.

[157] F. J. Giessibl, M. Herz, and J. Mannhart. Friction traced to the single atom. *Proc. Natl. Acad. Sci. USA*, 99(19):12006–12010, 2002.

[158] A. DiCarlo, M. R. Scheinfein, and R. V. Chamberlin. Magnetic force microscopy utilizing an ultrasensitive vertical cantilever geometry. *Appl. Phys. Lett.*, 61(17):2108–2110, 1992.

[159] K. Kim, Y. Seo, H. Jang, S. Chang, Hong M.-H., and W. Jhe. Shear-mode magnetic force microscopy with a quartz tuning fork in ambient conditions. *Nanotechnology*, 17(7):S201–S204, 2006.

[160] U. Gysin, S. Rast, M. Kisiel, C. Werle, and E. Meyer. Low temperature ultrahigh vacuum noncontact atomic force microscope in the pendulum geometry. *Rev. Sci. Instrum.*, 82(2):023705, 2011.

[161] L. Huang and C. Su. A torsional resonance mode afm for in-plane tip surface interactions. *Ultramicroscopy*, 100(34):277–285, 2004.

[162] T. Kasai, B. Bhushan, L. Huang, and C. Su. Topography and phase imaging using the torsional resonance mode. *Nanotechnology*, 15(7):731–742, 2004.

[163] A. Kaidatzis and J. M. García-Martín. Torsional resonance mode magnetic force microscopy: enabling higher lateral resolution magnetic imaging without topography-related effects. *Nanotechnology*, 24(16):165704, 2013.

[164] C. Su, K. L. Babcock, and L. Huang. Torsional resonance mode probe-based instrument and method, 2005. US Patent 6,945,099 B1.

[165] M. Reinstädtler, T. Kasai, U. Rabe, B. Bhushan, and W. Arnold. Imaging and measurement of elasticity and friction using the TRmode. *J. Phys. D Appl. Phys.*, 38(18):R269–R282, 2005.

[166] S. P. Jarvis, H. Yamada, K. Kobayashi, A. Toda, and H. Tokumoto. Normal and lateral force investigation using magnetically activated force sensors. *Appl. Surf. Sci.*, 157(4):314–319, 2000.

[167] J. Stirling. Optimal geometry for a quartz multipurpose spm sensor. *Beilstein J. Nanotechnol.*, 4:370–376, 2013.

[168] Y. M. Tseytlin. Kinetostatic model of spring constant ratios for an AFM cantilever with end extended mass. *Ultramicroscopy*, 110(2):126–129, 2010.

[169] J. Fujita, M. Ishida, T. Sakamoto, Y. Ochiai, T. Kaito, and S. Matsui. Observation and characteristics of mechanical vibration in three-dimensional nanostructures and pillars grown by focused ion beam chemical vapor deposition. *J. Vac. Sci. Technol. B*, 19(6):2834, 2001.

[170] P. Becker, H. Bettin, H.-U. Danzebrink, M. Gläser, U. Kuetgens, A. Nicolaus, D. Schiel, P. De Bièvre, S. Valkiers, and P. Taylor. Determination of the avogadro constant via the silicon route. *Metrologia*, 40(5):271–287, 2003.

[171] C. A. Clifford and M. P. Seah. The determination of atomic force microscope cantilever spring constants via dimensional methods for nanomechanical analysis. *Nanotechnology*, 16(9):1666–1680, 2005.

[172] J. Lübbe, L. Doering, and M. Reichling. Precise determination of force microscopy cantilever stiffness from dimensions and eigenfrequencies. *Meas. Sci. Technol.*, 23(4):045401, 2012.

[173] P. F. Carcia. Perpendicular magnetic anisotropy in Pd/Co and Pt/Co thin-film layered structures. *J. Appl. Phys.*, 63(10):5066, 1988.

[174] J. W. Knepper and F. Y. Yang. Oscillatory interlayer coupling in Co-Pt multilayers with perpendicular anisotropy. *Phys. Rev. B*, 71(22), 2005.

[175] M. A. Ruderman and C. Kittel. Indirect exchange coupling of nuclear magnetic moments by conduction electrons. *Phys. Rev.*, 96(1):99–102, 1954.

[176] T. Kasuya. A theory of metallic ferro- and antiferromagnetism on Zener's model. *Prog. Theor. Phys.*, 16(1):45–57, 1956.

[177] K. Yosida. Magnetic properties of Cu-Mn alloys. *Phys. Rev.*, 106(5):893–898, 1957.

[178] S. Hashimoto, Y. Ochiai, and K. Aso. Perpendicular magnetic anisotropy and magnetostriction of sputtered Co/Pd and Co/Pt multilayered films. *J. Appl. Phys.*, 66(10):4909, 1989.

[179] S. Vock. *Resolving Local Magnetization Structures by Quantitative Magnetic Force Microscopy.* PhD thesis, Technische Universität Dresden, 2014.

[180] J. E. Davies, O. Hellwig, E. E. Fullerton, G. Denbeaux, J. B. Kortright, and K. Liu. Magnetization reversal of Co/Pt multilayers: Microscopic origin of high-field magnetic irreversibility. *Phys. Rev. B*, 70(22), 2004.

[181] S. Vock, C. Hengst, M. Wolf, K. Tschulik, M. Uhlemann, Z. Sasvári, D. Makarov, O. G. Schmidt, L. Schultz, and V. Neu. Magnetic vortex observation in FeCo nanowires by quantitative magnetic force microscopy. *Appl. Phys. Lett.*, 105(17):172409, 2014.

[182] A. Hubert, W. Rave, and S.L. Tomlinson. Imaging magnetic charges with magnetic force microscopy. *Phys. Status Solidi B*, 204(2):817–828, 1997.

[183] Silvia Vock. Personal communication: 11/12/2013.

[184] C. F. Reiche, S. Vock, V. Neu, L. Schultz, B. Büchner, and T. Mühl. Bidirectional quantitative force gradient microscopy. *New J. Phys.*, 17(1):013014, 2015.

[185] I. Horcas, R. Fernández, J. M. Gómez-Rodríguez, J. Colchero, J. Gómez-Herrero, and A. M. Baro. WSXM: A software for scanning probe microscopy and a tool for nanotechnology. *Rev. Sci. Instrum.*, 78(1):013705, 2007.

[186] P. J. A. van Schendel, H. J. Hug, B. Stiefel, S. Martin, and H.-J. Güntherodt. A method for the calibration of magnetic force microscopy tips. *J. Appl. Phys.*, 88(1):435, 2000.

[187] S. Vock, Z. Sasvari, C. Bran, F. Rhein, U. Wolff, N. S. Kiselev, A. N. Bogdanov, L. Schultz, O. Hellwig, and V. Neu. Quantitative magnetic force microscopy study of the diameter evolution of bubble domains in a $(Co/Pd)_{80}$ multilayer. *IEEE Trans. Magn.*, 47(10):2352–2355, 2011.

[188] E. du Trémolet de Lacheisserie, D. Gignoux, and M. Schlenker, editors. *Magnetism: materials and applications*. Number 2 in Grenoble Sciences. Springer, Berlin, Heidelberg, 2005.

[189] C. P. Green and J. E. Sader. Torsional frequency response of cantilever beams immersed in viscous fluids with applications to the atomic force microscope. *J. Appl. Phys.*, 92(10):6262, 2002.

[190] J. R. Banerjee. Explicit frequency equation and mode shapes of a cantilever beam coupled in bending and torsion. *J. Sound Vibr.*, 224(2):267–281, 1999.

[191] M. Munz. Force calibration in lateral force microscopy: a review of the experimental methods. *J. Phys. D Appl. Phys.*, 43(6):063001, 2010.

[192] J. D. Parkin and G. Hähner. Calibration of the torsional and lateral spring constants of cantilever sensors. *Nanotechnology*, 25(22):225701, 2014.

[193] H. Czichos and M. Hennecke, editors. *Hüette: Das Ingenieurswissen*. Springer, 33th revised edition, 2008.

[194] D. F. Ogletree, R. W. Carpick, and M. Salmeron. Calibration of frictional forces in atomic force microscopy. *Rev. Sci. Instrum.*, 67(9):3298, 1996.

[195] C. P. Green, H. Lioe, J. P. Cleveland, R. Proksch, P. Mulvaney, and J. E. Sader. Normal and torsional spring constants of atomic force microscope cantilevers. *Rev. Sci. Instrum.*, 75(6):1988, 2004.

[196] C. F. Reiche, J. Körner, B. Büchner, and T. Mühl. Introduction of a co-resonant detection concept for mechanical oscillation-based sensors. *Nanotechnology*, 26(33):335501, 2015.

[197] J. W. Baldwin, Zalalutdinov M. K., Feygelson T., Pate B. B., Butler J. E., and Houston B. H. Nanocrystalline diamond resonator array for RF signal processing. *Diam. Relat. Mater.*, 15(1112):2061, 2006.

[198] A. A. Seshia, M. Palaniapan, T. A. Roessig, R. T. Howe, R. W. Gooch, T. R. Schimert, and S. Montague. A vacuum packaged surface micromachined resonant accelerometer. *J. Microelectromech. Syst.*, 11(6):784–793, 2002.

[199] B. Ilic, H. G. Craighead, S. Krylov, W. Senaratne, C. Ober, and P. Neuzil. Attogram detection using nanoelectromechanical oscillators. *J. Appl. Phys.*, 95(7):3694, 2004.

[200] Ü. Sökmen, A. Stranz, A. Waag, A. Ababneh, H. Seidel, U. Schmid, and E. Peiner. Evaluation of resonating si cantilevers sputter-deposited with AlN piezoelectric thin films for mass sensing applications. *J. Micromech. Microeng.*, 20(6):064007, 2010.

[201] C. L. Degen, M. Poggio, H. J. Mamin, C. T. Rettner, and D. Rugar. Nanoscale magnetic resonance imaging. *Proc. Natl. Acad. Sci. USA*, 106(5):1313, 2009.

[202] J. Chaste, A. Eichler, J. Moser, G. Ceballos, R. Rurali, and A. Bachtold. A nanomechanical mass sensor with yoctogram resolution. *Nat. Nanotech.*, 7(5):301, 2012.

[203] D. Garcia-Sanchez, A. San Paulo, M. Esplandiu, F. Perez-Murano, L. Forr, A. Aguasca, and A. Bachtold. Mechanical detection of carbon nanotube resonator vibrations. *Phys. Rev. Lett.*, 99(8):085501, 2007.

[204] S. Stapfner, L. Ost, D. Hunger, J. Reichel, I. Favero, and E. M. Weig. Cavity-enhanced optical detection of carbon nanotube Brownian motion. *Appl. Phys. Lett.*, 102(15):151910, 2013.

[205] S. D. Solares and G. Chawla. Dual frequency modulation with two cantilevers in series: a possible means to rapidly acquire tip-sample interaction force curves with dynamic afm. *Meas. Sci. Technol.*, 19(5):055502, 2008.

[206] M. Spletzer, A. Raman, A. Q. Wu, X. Xu, and R. Reifenberger. Ultrasensitive mass sensing using mode localization in coupled microcantilevers. *Appl. Phys. Lett.*, 88(25):254102, 2006.

[207] H. Pakdast and M. Lazzarino. Triple coupled cantilever systems for mass detection and localization. *Sensor. Actuat. A Phys.*, 175:127, 2012.

[208] E. Gil-Santos, D. Ramos, A. Jana, M. Calleja, A. Raman, and J. Tamayo. Mass sensing based on deterministic and stochastic responses of elastically coupled nanocantilevers. *Nano Lett.*, 9(12):4122, 2009.

[209] T. S. Biswas, J. Xu, X. Rojas, C. Doolin, A. Suhel, K. S. D. Beach, and J. P. Davis. Remote sensing in hybridized arrays of nanostrings. *Nano Lett.*, 14(5):2541, 2014.

[210] B. E. DeMartini, J. F. Rhoads, S. W. Shaw, and K. L. Turner. A single input-single output mass sensor based on a coupled array of microresonators. *Sensor. Actuat. A Phys.*, 137(1):147, 2007.

[211] R. Karabalin, M. Cross, and M. Roukes. Nonlinear dynamics and chaos in two coupled nanomechanical resonators. *Phys. Rev. B*, 79(16):165309, 2009.

[212] S.-B. Shim, M. Imboden, and P. Mohanty. Synchronized oscillation in coupled nanomechanical oscillators. *Science*, 316(5821):95, 2007.

[213] H. Okamoto, A. Gourgout, C.-Y. Chang, K. Onomitsu, I. Mahboob, E. Y. Chang, and H. Yamaguchi. Coherent phonon manipulation in coupled mechanical resonators. *Nat. Phys.*, 9(8):480–484, 2013.

[214] D. F. Wang, T. Itoh, T. Ikehara, and R. Maeda. Doubling flexural frequency response using synchronised oscillation in a micromechanically coupled oscillator system. *Micro. Nano. Lett.*, 7(8):717, 2012.

[215] T. Faust, J. Rieger, M. J. Seitner, P. Krenn, J. P. Kotthaus, and E. M. Weig. Nonadiabatic dynamics of two strongly coupled nanomechanical resonator modes. *Phys. Rev. Lett.*, 109(3), 2012.

[216] X. Li, R. Lin, and K. W. Leow. Performance-enhanced micro-machined resonant systems with two-degrees-of-freedom resonators. *J. Micromech. Microeng.*, 10(4):534–539, 2000.

[217] X. Li, T. Ono, R. Lin, and M. Esashi. Resonance enhancement of micromachined resonators with strong mechanical-coupling between two degrees of freedom. *Microelectronic Engineering*, 65(1-2):1–12, 2003.

[218] S. Gröblacher, K. Hammerer, M. R. Vanner, and M. Aspelmeyer. Observation of strong coupling between a micromechanical resonator and an optical cavity field. *Nature*, 460(7256):724–727, 2009.

[219] P. Rabl, P. Cappellaro, M. V. Gurudev Dutt, L. Jiang, J. R. Maze, and M. D. Lukin. Strong magnetic coupling between an electronic spin qubit and a mechanical resonator. *Phys. Rev. B*, 79(4), 2009.

[220] F. Torres, G. Abadal, J. Arcamone, J. Teva, J. Verd, A. Uranga, J. L. Lopéz, X. Borrisé, F. Pérez-Murano, and N. Barniol. Coupling resonant micro and nanocantilevers to improve mass responsivity by detectability product. *Intern. Solid State Sensor. Actuator. Microsys. Conf.*, pages 237–240, 2007.

[221] A. E. Miroshnichenko, S. Flach, and Y. S. Kivshar. Fano resonances in nanoscale structures. *Rev. Mod. Phys.*, 82:2257–2298, 2010.

[222] L. Novotny. Strong coupling, energy splitting, and level crossings: A classical perspective. *Am. J. Phys.*, 78(11):1199, 2010.

[223] H. Irretier. *Grundlagen der Schwingungstechnik*, volume 2. Vieweg, Braunschweig, Wiesbaden, 2001.

[224] C. F. Reiche, T. Mühl, and J. Körner. Verfahren und Vorrichtung zur Ermittlung von Kraftfeldern, Kraftfeldgradienten, Materialeigenschaften oder Massen mit einem System aus gekoppelten, schwingungsfähigen, balkenartigen Komponenten. Patent application. Submitted on 12/12/2014.

[225] G. Vidal-Álvarez, J. Agustí, F. Torres, G. Abadal, N. Barniol, J. Llobet, M. Sansa, M. Fernández-Regúlez, F. Pérez-Murano, Á. San Paulo, and O. Gottlieb. Top-down silicon microcantilever with coupled bottom-up silicon nanowire for enhanced mass resolution. *Nanotechnology*, 26(14):145502, 2015.

[226] M. H. Korayem, N. Ebrahimi, and S. Moloodi. Sensitivity analysis of carbon nanotube characteristics on dynamic behavior of tapping-mode atomic force microscopy. *Nanosci. Nanotech.*, 2(3):57–65, 2012.

[227] J. Chen, E. Louis, H. Wormeester, R. Harmsen, R. van de Kruijs, C. J. Lee, W. van Schaik, and F. Bijkerk. Carbon-induced extreme ultraviolet reflectance loss characterized using visible-light ellipsometry. *Meas. Sci. Technol.*, 22(10):105705, 2011.

[228] S. Dohn, W. Svendsen, A. Boisen, and O. Hansen. Mass and position determination of attached particles on cantilever based mass sensors. *Rev. Sci. Instr.*, 78(10):103303, 2007.

[229] A. Botman, M. Hesselberth, and J. J. L. Mulders. Investigation of morphological changes in platinum-containing nanostructures created by electron-beam-induced deposition. *J. Vac. Sci. Technol. B*, 26(6):2464–2467, 2008.

[230] A. Botman, J. J. L. Mulders, and C. W. Hagen. Creating pure nanostructures from electron-beam-induced deposition using purification techniques: a technology perspective. *Nanotechnology*, 20(37):372001, 2009.

[231] D. A. Dikin, X. Chen, W. Ding, G. Wagner, and R. S. Ruoff. Resonance vibration of amorphous SiO_2 nanowires driven by mechanical or electrical field excitation. *J. Appl. Phys.*, 93(1):226, 2003.

[232] E. Gil-Santos, D. Ramos, J. Martínez, M. Fernández-Regúlez, R. García, Á. San Paulo, M. Calleja, and Javier Tamayo. Nanomechanical mass sensing and stiffness spectrometry based on two-dimensional vibrations of resonant nanowires. *Nat. Nanotechnol.*, 5(9):641–645, 2010.

[233] F. R. Braakman, D. Cadeddu, G. Tütüncüoglu, F. Matteini, D. Rüffer, A. Fontcuberta i Morral, and M. Poggio. Nonlinear motion and mechanical mixing in as-grown GaAs nanowires. *Appl. Phys. Lett.*, 105(17):173111, 2014.

[234] Julia Körner. Personal communication.

[235] H.-J. Butt and M. Jaschke. Calculation of thermal noise in atomic force microscopy. *Nanotechnology*, 6(1):1–7, 1995.

[236] E. Majorana and Y. Ogawa. Mechanical thermal noise in coupled oscillators. *Phys. Lett. A*, 233(3):162–168, 1997.

[237] J. Körner, C. F. Reiche, B. Büchner, G. Gerlach, and T. Mühl. Application of a co-resonant sensor concept in cantilever magnetometry. Proc. IEEE 15th Int. Conf. Nanotech., 2015. in press.

[238] I. N. Bronstein, K. A. Semendjajew, G. Musiol, and H. Mühlig. *Taschenbuch der Mathematik*. Harri Deutsch, Frankfurt am Main, Thun, 4th revised edition, 1999.

Acknowledgments

First of all I want to express my gratitude to Prof. Dr. Bernd Büchner for giving me the opportunity to write my thesis on novel sensors for scanning force microscopy at the IFW Dresden. I would also like to thank him for his guidance and advice.

I also want to thank Prof. Dr. Georg Weiß for his interest in my work and for accepting to act as a referee.

I owe a dept of gratitude to Dr. Thomas Mühl for his supervision, encouragement and the many interesting discussions both related and unrelated to the work presented this thesis. I also want to thank him for proofreading the manuscript.

I am grateful to Julia Körner for the good collaboration and the numerous inspiring discussions on many different topics. I am further grateful for her opinion on the manuscript and the proofreading of this manuscript.

My thanks also goes to the many other former an current colleagues, who created a great working environment and helped me deepen my understanding of various physics-related topics. This includes Dr. Stefan Philippi, Frederik Klein, Dr. Robert Streubel, Clemens Gütter, Markus Gellesch and many more.

Furthermore, I want to thank all my collaborators for the good co-operation on many scientific projects. This includes Dr. Silvia Vock, Dr. Volker Neu, Dr. Tobias Schwarz and his colleagues from Tübingen as well as Dr. Jürgen Lindner and his group. A special thanks goes to Volker Neu and Tobias Schwarz for proofreading smaller parts of this manuscript.

I am grateful to Dr. Danny Baumann (IFW Dresden) and Dr. Timothy Ashworth (NanoScan AG) for providing insight into maintaining and repairing experimental equipment.

I want to also thank Dr. Thomas Gemming and Dr. Siegfried Menzel for the access to the focused ion beam equipment and their support with related questions.

Support with the administrative tasks at the institute by Kerstin Höllerer and Katja Schmiedel is gratefully acknowledged.

I further want to extend my thanks to Dr. Nicolai Urban for sharing his insight on language related questions and for proofreading large parts of this manuscript.

Last but not least I wish to thank my family, especially my parents and my brother for their ongoing support and understanding.